Assessing Military Benefits of S&T Investments in Micro Autonomous Systems Utilizing a Gedanken Experiment

Albert Sciarretta, Joseph Mait, Richard Chait, Elizabeth Redden, and Jordan Willcox

Center for Technology and National Security Policy

National Defense University

January 2011

The views expressed in this article are those of the authors and do not reflect the official policy or position of the National Defense University, the Department of Defense, or the U.S. Government. All information and sources for this paper were drawn from unclassified materials.

Albert Sciarretta is a Senior Research Fellow at the National Defense University (NDU) Center for Technology and National Security Policy (CTNSP) and president of Carbon Niter Sulfur (CNS) Technologies, Inc. He received MS degrees in operations research and mechanical engineering from Stanford University and a BS degree in general engineering from the U.S. Military Academy.

Joseph N. Mait, PhD, is a Senior Technical Researcher at the U.S. Army Research Laboratory (ARL). He is the Cooperative Agreement Manager of the Collaborative Technology Alliance (CTA) on Micro-Autonomous Systems and Technology (MAST). Dr. Mait received his PhD and MS degrees in electrical engineering from Georgia Tech, and his BS degree in electrical engineering from the University of Virginia.

Richard Chait, PhD, is a Distinguished Research Fellow at CTNSP. He was previously Chief Scientist, Army Material Command, and Director, Army Research and Laboratory Management. Dr. Chait received his PhD in solid state science from Syracuse University and a BS degree from Rensselaer Polytechnic Institute.

Elizabeth S. Redden, PhD, has been the Chief of the ARL Human Research and Engineering Directorate's (HRED's) Ft. Benning Field Element since 1982, focusing on human factors engineering for warfighting systems. She served as the Army Technology Objective Manager for Situational Understanding and currently serves as the U.S. National Lead for The Technical Cooperative Program (TTCP) Human Resources and Performance Group (HUM) 15 Land Panel. She received a PhD in organizational analysis from Auburn University, an MBA from Columbus State University, and MS and BS degrees in psychology from Valdosta State University.

Jordan Willcox served as a Research Assistant at CTNSP until the completion of his MA degree in Security Studies from Georgetown University in 2009. He is currently employed in social science and simulation-related research, analysis, and validation tasks as an employee of System of Systems Analytics, Inc. His prior experience in research, policy writing, and database design was performed at the Nixon Center and the Organization for American States.

Acknowledgements The authors wish to acknowledge Dr. Thomas Killion, Army Science and Technology Executive; Dr. Hans Binnendijk, Director of CTNSP; and Dr. Michael Baranick, Senior Research Fellow at CTNSP, for their support of this effort. We thank Mr. Dan Turner and Mr. Rodger Pettitt, HRED Field Element, Fort Benning, GA, for their technical support and for organizing this event. We thank Mr. Michael Kennedy, Robotics Lead, Maneuver Battle Lab, Fort Benning, GA, for his technical support and participation. We thank the members of the MAST CTA for their participation. Finally, and by far not least, we thank the Warfighters who participated in the event; especially Sergeant First Class (SFC) David J. Chappelle, SFC Troy A. Jensen, and SFC Rodney D. Pittman, who were instrumental in soliciting participants and providing the small unit urban warfare tutorials and training.

Executive Summary

In a Defense and Technology Paper (DTP) entitled "A Methodology for Assessing the Military Benefits of Science and Technology Investments,"[1] the National Defense University (NDU) Center for Technology and National Security Policy (CTNSP) presented a variety of approaches for deriving the return on investment – in terms of warfighting capabilities – for Army science and technology (S&T) efforts. As a follow-up to the methodology study that generated the DTP, the CTNSP wished to demonstrate parts of the methodology in the evaluation of an actual Army S&T effort. The Army Research Laboratory's (ARL's) Micro-Autonomous Systems and Technology (MAST) Collaborative Technology Alliance (CTA)[2] program was chosen to demonstrate the utility of the methodology because it offers significant future capabilities for our Army, provides a set of very robust present-day technical challenges, and offers a significant assessment challenge since it is focused on basic research.

The MAST CTA demonstration focused on the components of the methodology that relate to modeling and simulation. It was conducted as a *Gedanken Experiment* (thought experiment) to test hypotheses because the micro-autonomous robotic systems are only conceptual at this time. It explored the potential operational benefits of the MAST technologies by attempting to prove or disprove hypotheses using the thoughts of subject matter experts (SMEs) rather than experimental data derived from brassboard, breadboard, prototype, or fielded systems. Similar types of experiments have been carried out under names like *Technology Wargame*, *Army Transformation Wargame*, etc. Within these low fidelity simulations, participants provide SME feedback as they step through the use of the technologies in the planning and execution cycles of various warfighting mission scenarios, taking into account various topics, capabilities, constraints, and trade-offs. In demonstrating the applicability of the cost-benefit methodology, the Gedanken Experiment provided the MAST CTA Cooperative Agreement Manager (CAM) with Warfighter insights which proved to be useful for design considerations and future trade-off analyses. As hoped for in a cost-benefit analysis effort, it aided the MAST CTA CAM in developing his investment strategy.

The Gedanken Experiment activity was a three day event at Fort Benning, GA. The first day of the event was devoted to exposing the MAST CTA personnel (the CAM and MAST technical performers) to small unit operations (how dismounted infantry fire teams and squads clear rooms and buildings) and providing them an overview of the three scripted experiment scenarios. This exposure included hands-on experience in clearing rooms at a Fort Benning training site. During the second and third days, nine of the MAST CTA personnel and two separate groups of Warfighters participated in the Gedanken Experiment and follow-on survey activities in a classroom environment. For the experiment, the moderator used MS PowerPoint presentations of the scenarios to guide the participants through their discussions. At the end of each day, ARL Human Research and Engineering Directorate (HRED) Fort Benning Field Element personnel conducted a survey. An analysis of the discussions and survey data was conducted by the CTNSP personnel, the MAST CTA CAM, and HRED personnel.

---

[1] Albert Sciarretta, Richard Chait, Joseph Mait, and Jordan Willcox, *A Methodology for Assessing the Military Benefits of Science and Technology Investments*, Defense & Technology Paper 55 (Washington, D.C.: Center for Technology and National Security Policy, September 2008).
[2] A program description is available online at <http://www.arl.army.mil/www/default.cfm?Action=93&Page=332>.

The overall effort was a successful endeavor for the MAST CTA. The MAST CTA personnel gained much from their exposure to small unit urban operations, including the first day's classroom tutorial and the hands-on training experience at an urban dismounted infantry training site. Their involvement in the Gedanken Experiments on the second and third day:

- Exposed them to Warfighter operational thought processes and concerns
- Exposed them to Warfighter tactics, techniques, and procedures (TTP), which generated ideas on how MAST systems might be integrated into small unit urban operations
- Utilized their subject matter expertise to answer the technical/capability questions of the Warfighter participants

In addition, the Warfighter participants gained insights on what the Army's science and technology community is trying to do for them.

Many technical and operational topics are highlighted in this paper. The following were of most benefit to the MAST CTA CAM:

- The Warfighters' highest capability priorities were information fidelity (including sensing), mobility, and stealth. In this case, stealth is more of a camouflage or blending in with the surroundings and reducing noise; as opposed to reductions in visual, electromagnetic, and radar signatures.

- For most areas, the platoon level leaders and the squad level leaders agreed on uses and needs. There were some notable differences, especially in the area of information fidelity (platoon level leaders wanted more all the time), real-time feedback (platoon level leaders were willing to trade fidelity for real-time feedback), command and control (C2) of the systems (during mission execution, squad level leaders wanted hands off, while platoon level leaders wanted squads to maintain operational control), and overall control (squad level leaders wanted a hierarchy of control and maintenance at platoon level, while platoon level leaders saw all control and maintenance at the squad level).

- For TTP, novel ideas for deployment were discussed and concerns for C2 kept arising. With respect to C2, a need was identified for an easy-to-use Warfighter-MAST systems interface that supports Warfighter-machine interactions, enhances Warfighter situational awareness, and allows for changes in operational control during the execution of a mission.

- Additionally for C2, there is a need for scalable interfaces that provide differing views of C2 for platoon level leaders and squad level leaders, or the differing needs for levels of information fidelity during an operation. For example, platforms can best be exploited for intelligence gathering to aid in planning but use should be limited during actual combat. To accommodate these differences, multiple MAST designs may be needed along with scalable levels of autonomy among an ensemble of MAST systems of varying types, so that autonomy levels can be increased or decreased based upon the mission phase. For example, when the MAST technology is immature and autonomous behavior is not very predictable, it may be best to scale down autonomy during a firefight.

- Although lethal capabilities were discussed, there appeared to be more of an interest in non-lethal capabilities to support deception and diversion during offensive operations and to possibly reduce the number of civilian casualties, especially in hostage situations. For the MAST CTA CAM, this was an unforeseen beneficial warfighting capability that is potentially doable for micro autonomous systems.

The most important benefit for the MAST CTA CAM was his use of the Warfighters' perspectives to assess the benefits of design approaches. For example, platforms can be best exploited for intelligence gathering to aid in planning but should not be used when the potential for combat operations is high. For cultural reasons (i.e., military culture) and the immaturity of the technology, Warfighters should not be teamed with autonomous platforms during a firefight. This single perspective has helped MAST CTA CAM keep the researchers focused on a well defined design approach.

The MAST CTA CAM acknowledged that not only did the Gedanken Experiment output influence his leadership of MAST, but as ARL defines its strategic vision in autonomous platforms of all sizes and shapes, it has also provided him supporting information to debate with other researchers in the lab about the reasonableness of promoting room clearing by autonomous platforms. He believes he has successfully argued that this is not a good application to consider, at least not until the technology has matured and Warfighters trust the capabilities of such platforms.

The overall experience was very beneficial to all participants in the exercise. It is highly recommended that the MAST CTA keep a close working relationship with Warfighters, especially those involved with developing TTP and assessing Warfighter-machine interfaces at Fort Benning, GA.

Table of Contents

Background

In a Defense and Technology Paper (DTP) entitled "A Methodology for Assessing the Military Benefits of Science and Technology Investments,"[3] the National Defense University (NDU) Center for Technology and National Security Policy (CTNSP) presented a variety of approaches for deriving the return on investment – in terms of warfighting capabilities – for Army science and technology (S&T) efforts. As a follow-up to the methodology study that generated the DTP, the CTNSP wished to demonstrate parts of the methodology in the evaluation of an actual Army S&T effort. The Army Research Laboratory's (ARL's) Micro-Autonomous Systems and Technology (MAST) Collaborative Technology Alliance (CTA)[4] program was chosen to demonstrate the utility of the methodology because it offers significant future capabilities for our Army, provides a set of very robust present-day technical challenges, and offers a significant assessment challenge since it is focused on basic research.

The MAST CTA demonstration focused on the components of the methodology that relate to modeling and simulation. It was conducted as a *Gedanken Experiment* (thought experiment) to test hypotheses because the micro-autonomous robotic systems are only conceptual at this time. It explored the potential operational benefits of the MAST technologies by attempting to prove or disprove hypotheses using the thoughts of subject matter experts (SMEs) rather than experimental data derived from brassboard, breadboard, prototype, or fielded systems. Similar types of experiments have been carried out under names like *Technology Wargame*, *Army Transformation Wargame,* etc. Within these low fidelity simulations, participants provide SME feedback as they step through the use of the technologies in the planning and execution cycles of various warfighting mission scenarios, taking into account various topics, capabilities, constraints, and trade-offs. In demonstrating the applicability of the cost-benefit methodology, the Gedanken Experiment provided the MAST CTA Cooperative Agreement Manager (CAM) with Warfighter insights which proved to be useful for design considerations and future trade-off analyses. As hoped for in a cost-benefit analysis effort, it aided the MAST CTA CAM in developing his investment strategy.

Importance of Robotic Systems

The Enduring Freedom and Iraqi Freedom Operations have demonstrated the value of robotic platforms, both aerial and ground, that are remotely tele-operated. Robotic platforms extend the Warfighter's senses and reach and have been used as sensors, communication devices, and, in some instances, weapons platforms. Within complex terrain, like caves, mountains, or urban environments, these platforms provide operational capabilities to the Warfighter that would otherwise be costly, impossible, or deadly to achieve with humans. Future enhancements to warfighting capabilities require a reduction in platform size and the cohesive operation of multiple platforms that operate with little or no direct human supervision and can support personnel operating in a variety of dangerous environments.

---

[3] Albert Sciarretta, Richard Chait, Joseph Mait, and Jordan Willcox, *A Methodology for Assessing the Military Benefits of Science and Technology Investments*, Defense & Technology Paper 55 (Washington, D.C.: Center for Technology and National Security Policy, September 2008).

[4] A program description is available online at <http://www.arl.army.mil/www/default.cfm?Action=93&Page=332>.

The MAST CTA

The Army established the MAST CTA in February 2008 to address these future capabilities for autonomous platforms at its corporate laboratory, the Army Research Laboratory. The objective of the MAST CTA is to enhance tactical situational awareness in urban and complex terrain by enabling the autonomous operation of a collaborative ensemble of multifunctional, mobile micro-systems. See Figure 1 for a conceptual view of an operational use (detecting explosives in a cave) of MAST systems teamed with a small unmanned ground vehicle (SUGV).

One goal of the MAST CTA is to advance fundamental S&T in several key areas of robotics including small-scale aeromechanics and ambulation; propulsion; sensing, processing, and communications; navigation and control; micro devices and integration; platform packaging; and systems architectures. The approach is to consider the interplay among all elements, as opposed to each element independently.

Figure 1. Air and ground MAST systems working with a relatively larger SUGV (right hand side of the picture)

To achieve this goal, radical design and engineering methodologies are required in which system-level performance, maneuvering, and functional adaptability are emphasized over the optimization of individual functions. The Gedanken Experiment proved to be a beneficial tool for guiding the design of actual platforms.

Functional and Operational Capabilities and Constraints

As stated, the vision of the MAST CTA is to enhance tactical situational awareness in urban and complex terrain. The impact and interplay among conflicting requirements on the MAST technical issues are so complex that investigating a single issue in isolation of the others will not generate an efficient and operationally effective ensemble of micro systems. For example, solutions for processing, communications, and mobility that are satisfactory for large systems do not scale when platforms are reduced to the size considered in the MAST CTA (i.e., palm-sized or smaller). Considering that platform size and weight limit available power over the duration of a mission, the largest percentage of available power is utilized for mobility. The limited power, in turn, constrains the bandwidth of intra-platform communications (i.e., between sensors and processors and between processors and transmitters), as well as the bandwidth of inter-platform communications. Limited communications further impact the ability of the micro-robotic systems to collectively sense, understand, and respond coherently as a group. The Gedanken Experiment technique provided a potential methodology to address design trade-offs such as these.

For example, one scenario considered for the potential application of small platforms included room-by-room searches, with MAST systems moving autonomously with a dismounted infantry squad. The Warfighters desired capabilities that would allow the MAST systems to provide indications of the presence of opposing force (OPFOR) personnel, threat munitions, or both. They stated that either an infrared image or a simple graphic indicator would be adequate. They clarified that the duration of operations (i.e., flight) in each room may be minutes; and to clear the building could take a half hour or more. Thus, sensing and duration of flight are critical functions for these platforms.

However, further analysis is still needed to determine the sensing fidelity and resolution, and the minimum flight duration required to ensure an effective mission. Sensors that can generate high fidelity, high resolution information require more power than low performance sensors and, as mentioned above, power is a limited resource. Thus, a trade-off between sensor performance and platform power is driven by mission effectiveness. Other system design considerations include maneuverability, obstacle avoidance, degree of human-system interface, and whether or not images will be transmitted in real time or viewed upon return of the micro autonomous system to the operator.

The Gedanken Experiment effort addressed these capability and constraint issues, and the findings were beneficial to the MAST CTA technical personnel.

Conduct of the Gedanken Experiment

In the early phases of the MAST CTA, technical personnel were asked to consider the capabilities required to achieve tactical situational awareness in urban and complex terrain in three different scenarios: small unit search within an intact building for OPFOR threats, perimeter defense of a special operations unit, and small unit search of a demolished building or a cave for potential threats. The Gedanken Experiment addressed three closely related scenarios that covered platoon-level urban operations:

- Offensive operation (see Appendix A). The scenario begins with the platoon planning the operation. The dismounted infantry platoon is tasked to clear two separate buildings, which are known meeting locations for the local insurgent forces. The platoon initiates the action by conducting an assault from the southwest corner of a town to seize Objectives (OBJs) Homer (building A) and Bart (building B). Support-by-fire positions (SBFs) are established to cover the entrances to the buildings and to prevent a counterattack by insurgents from OBJ Bart while OBJ Homer is being taken. The platoon is asked to use its organic SUGVs to establish observation points (OPs) to cover the rear exits of OBJ Bart. Initial reconnaissance of buildings and rooms is accomplished using SUGVs and smaller MAST systems. The primary task during this clearing operation is to identify and neutralize (capture or kill) all insurgents located within the objectives, while minimizing the number of civilian casualties.

- Defensive operation (see Appendix B). This is a very brief scenario. The dismounted infantry platoon conducts the offensive operation discussed above. Once OBJ Homer is cleared, the platoon receives a report that an additional ten to twenty insurgents are moving into the area. The company commander orders the platoon to take up defensive positions and await further orders.

– Stabilization operation (see Appendix C). An improvised explosive device (IED) explosion occurs in a large city and snipers are shooting at emergency responders. A reaction force (a U.S. infantry company augmented with a platoon made up of local nationals) is sent to secure the area and neutralize the snipers. Along the way the reaction force has to identify and deal with obstacles (crowded market, protestors, etc.). Once at the IED scene, the reaction force discovers that in addition to the snipers, local civilians may have been abducted and are being detained by insurgents. The reaction force receives additional orders to work with local police forces to verify the abduction, then locate and free the civilians.

The topics and approaches were designed to highlight capability needs and identify technical and operational constraints.

Within the Gedanken Experiment, Warfighters assisted in defining the needs and uses for MAST systems – i.e., system design needs; tactics, techniques, and procedures (TTP). As discussed in DTP 55, follow-on, higher fidelity modeling and simulation research and experimentation could further refine these findings. Following each Gedanken Experiment, the participants were also surveyed to solicit their prioritization of capability needs for MAST systems. This prioritization was useful in determining the value of benefits and formulating MAST design approaches.

The ARL Human Research and Engineering Directorate (HRED) Fort Benning Field Element and the U.S. Army Maneuver Center[5] Battle Laboratory assisted with coordinating the event at Fort Benning, GA, reviewing the scenarios, and providing warfighting subject matter expertise when needed. Additionally, they assisted with roundtable discussion. Finally, the HRED team designed, conducted, and consolidated the information acquired from surveys provided to the experiment participants at the end of each day. The consolidated feedback is in Appendices D and E.

Concept of Operations

The Gedanken Experiment activity was a three day event at Fort Benning, GA. The first day of the event was devoted to exposing the MAST CTA personnel to small unit operations (how dismounted infantry fire teams and squads clear rooms and buildings) and providing them an overview of the three scripted experiment scenarios. The second and third days were devoted to the conduct of the experiment, the conduct of surveys, and end-of-the-day general and technical discussions.

Figure 2. Tutorial by SFC David J. Chappelle

---

[5] Formerly the U.S. Army Infantry Center. The name changed to accommodate the transition and integration of the U.S. Army Armor School to Fort Benning, GA. The U.S. Armor School was at Fort Knox, KY.

There were nine civilians from the MAST CTA effort, including representatives from ARL, Georgia Tech, the University of California at Berkeley, the University of Maryland, the University of Michigan, the University of New Mexico, and the University of Pennsylvania. On the first day, they received a classroom tutorial on small unit urban operations (Figure 2) as well as a trip to a small unit urban training site. At the site, they were divided into infantry fire teams of four personnel; each person was provided a helmet, body armor, and a rifle; and they were asked to clear rooms in training structures (Figure 3).

Figure 3. Two MAST CTA personnel beginning to clear a room

Later, back in a classroom, the moderator for the Gedanken Experiment discussions conducted a walk-through of the scenarios, to provide the MAST CTA personnel insights on TTP and familiarize them with military terminology that would be discussed in the scenarios or possibly brought up by the Warfighter participants. The reasons for including the MAST CTA personnel as participants were:

- Expose them to Warfighter operational environments, thought processes, and concerns
- Expose them to Warfighter TTP to generate ideas on how MAST systems might be integrated into small unit urban operations
- Utilize their subject matter expertise during the Gedanken Experiment to answer the technical and capability questions of the Warfighter participants

The second and third days of the event focused on the Gedanken Experiments. Each day began with the CAM of the MAST CTA providing an overview of the MAST technology efforts to the participants. The HRED personnel had solicited Warfighter participants who had experience with urban operations in Iraq or Afghanistan, and some experience with or knowledge of the use of robots in small unit operations. Participants for the second day included seven Warfighters with recent Squad Leader[6] and Fire Team Leader[7] experience, and the nine MAST CTA civilians. On the third day, there were eight Warfighters with Platoon Leader[8] and Platoon Sergeant[9] experience and the same nine civilians. Detailed demographic data for all participants can be found in Appendix D.

---

[6] Normally a staff sergeant and leads two Fire Teams, see Table 1, Appendix A.
[7] Normally a sergeant and leads three Warfighters, see Table 1, Appendix A.
[8] Normally an officer with the rank of second lieutenant and leads the entire platoon, see Table 1, Appendix A.
[9] Normally a sergeant first class and assists the platoon leader with leading the whole platoon, see Table 1, Appendix A.

During the experiment, the moderator discussed the offensive, defensive, and stability operations scenarios, asking general and scenario specific questions, and solicited input on TTP, the importance of envisioned design characteristics, and other desired capabilities. These deliberations included a step-by-step breakdown of each scenario and how the MAST CTA technologies might be used. Following the scenario discussions, the participants completed surveys which asked them to rate the importance of various MAST system capabilities and functions. The detailed survey output is in Appendix E and is discussed in the "Findings" section of this document. Each day ended with final operations-related discussions with the Warfighters, followed by technical discussions without the Warfighters.

During the second and third days, time was also allotted for Fort Benning HRED and Battle Lab personnel to discuss their recent research efforts related to robotic systems with the MAST CTA personnel. These discussions did not include the Warfighters.

The Gedanken Experiment and the survey activities were held in a classroom environment. In the classroom, the moderator used MS PowerPoint presentations to guide the participants through their discussions. During the Gedanken Experiment, all those in the classroom other than the moderator and the participants were considered observers.

Procedure Rules

At the beginning of each day the moderator discussed a list of rules for conducting discussions. These rules recognized the importance of academic freedom in speech – so all discussion points were non-attribution. In addition, these rules helped control discussions, focused on the views of the participants, and limited the amount of observer participation.

Data Collection

During the Gedanken Experiment, data was collected in three primary forms:
- Notes from general discussions. These notes addressed discussions immediately before and after the scenarios, as well as discussions during presentations by the Director, MAST CTA; the Chief, HRED Fort Benning Field Element; and the representative from the Maneuver Center Battle Lab
- Notes covering scenario discussions; specifically the answers to the general questions in the "General Questions" section and the scenario specific questions, as in Appendix A
- Surveys at the end of each day

In addition to notes taken by the identified note takers, notes were also collected from participants and observers who voluntarily provided them.

General Questions

Before each scenario, participants and observers were asked if they had any general questions they wished to address before initiating the scenarios.

During Each Scenario

After each step of a scenario mission, the participants were asked the following general questions:

- *How should MAST systems be employed?*
  - *When*
  - *Outside/in buildings*
  - *Special TTP*
- *Presentation of intelligence*
  - *Real time, delayed, etc.*
  - *Visual, audible, etc.*
  - *Fidelity of information (image, detectability, location, etc.)*
- *Any difference for day versus night operations?*
- *How much situational awareness (SA) is needed?*
  - *Location of MAST systems*
  - *Orientation of MAST systems*
  - *Location of items of interest*
  - *Identification of items of interest*
  - *Understanding of activity being viewed*
- *How should the MAST systems be controlled?*
- *What is important?*
  - *Duration of flight*
  - *Ability to transmit information*
  - *Ability to return with downloadable information*
  - *Type of information (humans, inanimate objects, military equipment, etc.)*
  - *Importance of sensors (visual, II, thermal, magnetic, seismic, acoustic, etc.)*
  - *Fidelity of information*
  - *Warfighter-Machine Interface (how much autonomy)*
  - *Obstacle avoidance*
  - *Fly, walk, or both*
  - *Crawl up walls*
  - *Stealth (type: visual, audible, electromagnetic)*

After Each Scenario

After a scenario, the participants and observers were asked the following questions:
- For this specific scenario, what are the most important capabilities provided by the MAST systems?
- What capabilities are not important?
- What TTP need to be established?
- Who should control the MAST systems?
- Other comments?

Surveys

Each day of the Gedanken Experiment, at the end of the last scenario, the participants were given two surveys to fill out. As seen in the Appendices D and E, the responders are in three groups:
- Group A: eight platoon leaders and platoon sergeants (second Gedanken Experiment)
- Group B: seven squad leaders and fire team leaders (first Gedanken Experiment)
- Group C: nine civilians who participated in both experiments. [Note: The nine civilians were given the surveys only once.]

The first was a demographic survey, asking about military occupational specialties[10] (MOSs), rank, current duty position, time in service, military education, experience in combat, experience with robotic systems, computer experience, and assessment of infantry skills. A compilation of the information is in Appendix D.

The second survey, in Appendix E, is a compilation of the information gathered from questions focused on MAST systems. It was hoped that after having almost a full day of discussions about MAST systems and how they could be used in various scenarios, the informed feedback of the participants could be used to better prioritize the importance of MAST system capabilities and functions.

Copies of the individual surveys are not included in this paper. However, the consolidated output can be found in Appendix E. The MAST system survey had three major components. In the first component (Section A), the participants were asked to use a scale from 1 (No Experience) to 7 (Expert) to rate the importance of levels of ability for the following ten capability areas:
- Ability to Sense
- Information Fidelity
- Information Latency
- Single Task/Mission Duration
- Total Operational Time Before Recharge
- Mobility
- System Command and Control (C2) Echelon
- Level of Autonomy
- Stealth
- Physical Specification

For example, for "Information Latency," abilities were associated with various levels of transmission latency as well as an ability to return with downloadable information. As seen on the third page of Appendix A, a few clarifying comments were provided with this feedback.

In the second component (Section B), as seen on the fourth page of Appendix A, participants were asked to rank in order from 10 (best) to 1 (worst) the same 10 capabilities identified in the first component. No ties were allowed.

In the final component (Sections C through G), participants were asked to provide textual comments on the following topics:
- What TTP need to be established for use of MAST systems?
- List three things you liked about the concept of MAST systems.
- List three things you believe can be improved in the MAST system design/capabilities.

---

[10] Military personnel are categorized in their assigned job by a military occupational specialty (MOS). Each MOS is labeled with a short alphanumerical code, which consists of a two-digit number appended by letter. Related MOSs are grouped together by Career Management Fields (CMF). For example, an enlisted Soldier with MOS 11B works as an infantryman, and is part of CMF 11 (the CMF for infantry). An infantry officer is MOS 11A.

- List up to three things you would like the MAST systems to do that were not discussed.
- Additional comments.

The comments associated with Sections C through G are listed with their appropriate topics in the fourth through ninth pages of Appendix A.

Findings

The Gedanken Experiment proved to be of great benefit to the CAM of the MAST CTA and his personnel. As stated by the CAM:

> *Not only has this influenced my leadership of MAST, but as ARL defines its strategic vision in autonomous platforms of all sizes and shapes, I have debated with other researchers in the lab about the reasonableness of promoting room clearing by autonomous platforms. I believe I have argued the point successfully that this is NOT a good application to consider, at least not until the technology has matured and soldiers trust the capabilities of such platforms.*

Much information was gained in terms of design needs, capability needs, potential benefits, and concerns about misuse. In addition, the exposure of the MAST personnel to warfighting operations also provided a very useful "early user" interaction with materiel developers. Rather than provide a very tedious listing of all deliberations during the scenario discussions, this section will discuss the most interesting information.

Exposure of MAST Personnel to Warfighting Operations

As previously discussed, the MAST personnel were exposed to warfighting operations prior to the Gedanken Experiment in the following mediums:
- Classroom tutorial on small unit urban operations, specifically the clearing of rooms and buildings
- On site, hands-on training experience of donning a helmet, body armor, and a rifle; and physically clearing rooms in Fort Benning training structures
- Classroom reviews of the Gedanken Experiment scenarios and basic introduction to TTP and military terminology

These three mediums significantly enhanced their understanding of potential uses of MAST systems in small unit urban operations. Additionally, it prepared them for their "participation" in the follow-on Gedanken Experiments over the next two days. They might have received significantly less benefit from their participation in the Gedanken Experiments without this one-day preparatory education.

This preparatory experience exposed participants to small unit urban operations, especially in terms of the speed of operations, complexity of movements, competing activities that may take the Warfighter's attention away from the MAST systems, limited amount of Warfighter-system interface time, and an early awareness of the potential significance of various functional capabilities. Most importantly, during the Gedanken Experiments, it provided them a level of military awareness that validated them as participants alongside the attending Warfighters.

Scenario Discussions

This discussion of observations and findings from the Gedanken Experiments is in no particular order or priority. The information is believed to be of value to the development of the MAST systems. The discussions primarily focus on the Warfighter feedback. The civilian participants (i.e., members of the MAST CTA and NDU) asked many questions of the Warfighters, but they themselves did not provide information discussed in this section. The civilian participants' feedback is addressed in the next section, which assesses the survey responses.

Differing Perspectives. In general, both groups were relatively conservative in identifying capability needs; however, platoon leaders and platoon sergeants were more conservative, especially when it came to the functions for which squad leaders and platoon leaders are responsible. Squad leaders were more imaginative (i.e., providing the MAST systems flash-bang capabilities for noisy diversionary activities in another part of a building, providing MAST systems non-lethal capabilities to neutralize insurgents while minimizing civilian casualties) and would be willing to take on more risk with respect to a function. Platoon leaders, given their responsibility for the entire mission, were less prone to venture outside doctrine without being prodded.

Impact on Momentum. Squad leaders and fire team leaders were mostly concerned about maintaining momentum (more commonly called operations tempo (OPTEMPO)[11] among military personnel) once an operation had begun. They did not want any diversions and had no desire to control platforms or look at displays when it came to kicking down doors, clearing rooms, and getting into firefights. As a side note, caution should be taken with this perception about OPTEMPO by Warfighters who have not had the opportunity to use detailed actionable information. One of the findings in an Office of the Secretary of Defense (OSD)-sponsored Smart Sensor Web (SSW) experiment at Fort Benning, in 2002, was that dismounted infantry platoons (blue forces) were willing to give up OPTEMPO in exchange for "actionable information"[12] provided by the SSW system.[13] Prior to the SSW experiment, it was believed that speed of action (high rate OPTEMPO) would compensate for lack of situational awareness[14] about adversaries (red forces), thus hopefully catching the red forces off-guard before they can react. It was observed and discussed in SSW after action reviews that when red force information was available, the blue forces could slow down and use their enhanced situational awareness to execute deliberate plans and enhance the survivability of the blue force.

---

[11] Tempo is a musical term meaning the speed at which a piece is played. OPTEMPO is short for "operations tempo" which is accepted among military personnel as a measure of the pace of an operation or operations.

[12] Actionable information is information that can be acted upon, leads to action, or starts a chain of events and reaction. Actionable information provides an initial basis point for hypothesis building. It is the strongest building block of the decision-making process.

[13] The SSW system provided the platoon leader, platoon sergeant, each squad leader, and each fire team leader a wearable computer with touch-screen wrist display. The computer was linked to a network of sensors (non-imaging, imaging, and streaming video) and databases (maps, room displays) of an urban environment (the Fort Benning McKenna Military Operations on Urban Terrain (MOUT) site). Enemy actions triggered sensors which provided real-time feeds of information to the SSW system users. Two of the authors, Mr. Albert Sciarretta and Dr. Elizabeth Redden, played key roles in the design, execution, and assessment of findings of the SSW experiment.

[14] A textbook definition of situational awareness is the perception of environmental elements within a volume of time and space, the comprehension of their meaning, and the projection of their status in the near future. In the Gedanken Experiment, the participants wanted a better understanding of the location of civilians, adversaries, and weapons, as well as the actions and possible intentions of adversaries.

Surveillance Needs. All platoon personnel wanted a perching capability (i.e., on the roof of a building or on a window sill) to provide persistent surveillance for 24–48 hours prior to operation. Soldiers felt that persistent surveillance and reconnaissance for mission planning support were more important capabilities than making the MAST systems lethal or even providing real-time information during the mission execution, including firefights.

For deployment, the Warfighters suggested a much simpler means for deployment than had been considered by the MAST CTA. For example, one suggestion was dropping the MAST systems while on patrol two days prior to a known operation. Another suggestion was to use them as removable hood ornaments on military vehicles, which provides Warfighters with capabilities while on mounted and dismounted operations.

Surveillance data, especially imagery for mapping rooms and identifying people, was desired for planning, execution, and after action reviews. In addition, the Warfighters provided many applications for acoustic sensor data. For example, they wanted both a portable and mobile version of an acoustic sniper detection system. Another suggested application was for acoustic sensing and guidance, in which MAST platforms could lead platoon members or other platforms to noisy rooms or provide access to OPFOR conversations. For this latter capability to be useful, though, Warfighters felt it required an organic translation capability or at least ability to transmit to a receiving station had the required translators.

With regard to surveillance data, the need for information fidelity was discussed. Figure 4 is a very simplistic depiction of the desires for information fidelity by both the platoon level leadership (platoon leaders and platoon sergeants) and the squad level leadership (squad leaders and fire team leaders) during three stages of an operation: mission planning, initiation of actions, and mission execution. Both the platoon level and squad level leaders wanted high fidelity information during the planning phase of a mission. Once the mission is executed, however, they were willing to accept a lower level of fidelity. That is, they want to know as much as possible about a situation prior to execution but, once the mission is under way, the simpler the sensory information, the better, i.e., instead of streaming video of the inside of a room, a simple alarm that merely indicates the presence of a human would be acceptable. During all phases of an operation, the platoon level leaders wanted more information and they wanted it in real time. They were willing to trade fidelity for real-time collection.

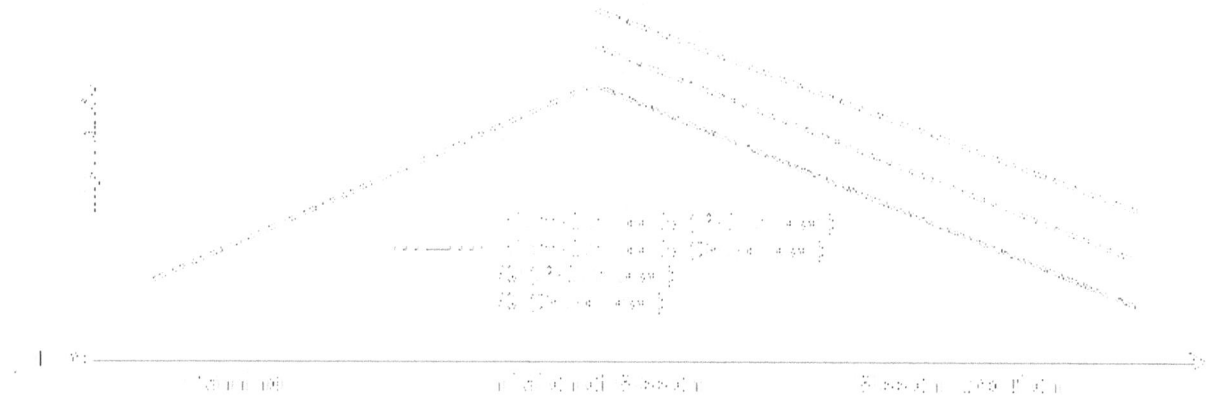

Figure 4. Comparison of views relative to information fidelity and C2

Command and Control. From a C2 perspective, as depicted in Figure 4, the views of the platoon level and squad level leaders was quite different. Squad level leaders wanted control of the MAST systems prior to execution but, during execution, did not want to be bothered with C2 issues. Momentum was their key concern. On the other hand, platoon level leaders wanted MAST systems to perform autonomous surveillance but, during mission execution, they felt human control of the platforms was essential to respond to fast moving chaotic events. The platoon level leaders also felt that squad leaders should be the ones in control, giving them an added advantage. However, this contradicts the squad level leaders' desire to have no distractions during room clearing operations, especially with regard to C2 of the MAST systems. This distinction highlights the need for tailored performance and tailored information flows to platoon level and squad level leaders. Also, there should be some consideration here about the relationship of OPTEMPO and gaining situational awareness, as described earlier with regard to the findings of the OSD SSW experiment. Perhaps the squad level leaders might change their opinions in a more advanced experiment with prototype systems.

In the Gedanken Experiment, all platoon personnel indicated a strong desire to keep assets and information organic with limited links back to company or higher level of command. They felt doing so would limit micromanagement. In addition, for C2 and maintenance ("care and feeding"), squad leaders and fire team leaders wanted an individual at the platoon level (i.e., a robotics non-commissioned officer (NCO) equivalent to a squad leader in rank) to maintain and control the MAST systems assigned to the platoon. This robotics NCO may come to fruition with current plans to provide SUGVs, perhaps as many as three, to infantry platoons. However, it is unclear if the robotics NCO will be able to single-handedly maintain and control the SUGVs as well as a large number of MAST systems (unknown at this time, but may be in the 10's for each squad). The squad leaders and fire team leaders believed that if the robotics NCO controlled and maintained the MAST systems, then for conduct of a mission, the MAST systems could be placed under operational control (OPCON)[15] of the squad leaders.

---

[15] Operational control (OPCON) allows for the command and control of subordinate forces, but does not require their care, feeding, administration, discipline, internal organization, or training. In this case, the robotics NCO would provide the maintenance and other administrative needs of the MAST systems.

Stealth. For stealth, the participants were more interested in camouflage, rather than dampening noise or creating visual transparencies. Warfighters believed the MAST systems should be like transformers (the toys). When perched, they should look like a rock, part of a roof or wall, or a local bug, not a piece of U.S. military equipment. When mobility is required, inanimate objects should transform into a bug.

Survivability. Survivability of the MAST systems was also discussed. In addition to physical survivability, there was also a concern for operating in an environment with electromagnetic interference whether that interference came from intentional adversary jammers, unintentional local equipment, or from friendly force systems (especially, counter IED systems).

Non-Lethal Capabilities. Non-lethal capabilities were also of interest to the participants; particularly, flash-bang or electric shock. The Warfighters were intrigued by the ability to use this non-lethal capability. For deception or diversion, the Warfighters wanted to send the non-lethal equipped MASTs to a secondary target (i.e., to a room that is not the primary target) to make noise or create havoc, thus making an adversary believe the secondary target is the primary objective of the mission. They also believed that the use of an electric shock capability could minimize casualties among innocent civilians, especially in hostage situations.

Ability to Learn. Learning was a unique capability squad level leaders wanted the MAST systems to have. Given that their troops will need to train with the platforms, squad level leaders felt the ability to recognize human behaviors would reduce the burden on troops. In addition, squad level leaders wanted the capability for learned behaviors from one set of MAST systems to be "pushed" to all platforms in a manner similar to how commercial software companies push updates. In this way, learning is transmitted throughout the MAST collective.

Warfighter-Machine Interfaces. There were many discussions among all participants about the ability of the Warfighters to interface with the MAST systems. C2 has been addressed above in terms of OPCON, but there is still an issue of how one would actually interact with single or multiple MAST systems. Also, Warfighters were concerned that when many MAST systems are employed in a mission area, it would be difficult to know which MAST is which and which camera is providing the information. This drives a need for MAST systems to have some capability to self locate and requires a greater need to present that self-location information, as well as sensor data, to operators in a way that it enhances situational awareness and does not reduce mission effectiveness.

Additional Observations

- In general, the Warfighters were able to comprehend the advantages of small and intelligent platforms. However, it appeared to the observers in the room that the advantages of a collaborative ensemble were harder for the Warfighters to appreciate. This may be due to the fact that the discussions of collaboration were at a more detailed technical level. For example, it was discussed that a collaborative ensemble of platforms can provide various geometric perspectives that allow the MAST systems to triangulate on acoustic and electro-magnetic radiation (visible, thermal, or radio frequency). The

Warfighters were less interested in how this is accomplished technically than they were in the function of locating, for example, acoustic sources (i.e., conversations or gun shots).

- Both groups of leaders felt that small unmanned ground vehicles (i.e., the current SUGV) may be disadvantaged in offensive operations. They felt the platforms were too large and noisy, which might cause the platoon to lose its element of surprise and may have some mobility issues. They preferred smaller, stealthier, more mobile platforms, and the MAST systems appeared to meet their preference.

Most of the discussion time in the experiment was focused on the offensive operation (the first scenario). The Warfighters spent a relatively small amount of time discussing applications in other operations.

- Defensive Operations. Use the MAST systems in perimeter security, as early warning devices (trip wires). Perimeter security could also be useful in offensive operations.

- Stabilization and Reconstruction Operations. Use the MAST systems for situational awareness (i.e., to look for a 2nd IED triggerman after first responders arrive on the scene of an initial EID explosion). Again, they may be useful in perimeter security, though the multitude of civilians and emergency responders might create a significant amount of clutter for MAST sensors and processors.

- Caves and Underground Tunnels. Use MAST systems to keep Warfighters out of harm's way. This drives a need for MAST system operations in the dark.

- Medical Operations. Squad level leaders, in particular, wanted to use the MAST systems to provide triage (i.e., an autonomous Star Trek medical *tricorder*[16]), whereas platoon level leaders wanted the platforms to provide security for medics. The platoon level leaders were less willing to use MAST systems in support of medical operations; they preferred human intervention.

Assessment of Survey Information

The following is a discussion of the survey information in Appendix E, Sections A through G. The format for the survey is discussed in the "Surveys" section.

Section A. In a review of the consolidated feedback for Section A, the three groups of responders were numerically close for most of the survey responses. Of note is how well the nine civilians (Group C), primarily MAST CTA personnel, aligned themselves with the two groups of Warfighters (Groups A (Platoon Leaders and Sergeants) and B (Squad Leaders and Fire Team Leaders)). This was mostly true in responses in Section A, which sought feedback on the importance of ten capability areas: ability to sense, information fidelity, information latency,

---

[16] A multifunction, handheld sensing, processing, and storage device used for scanning, data analysis, and recording data.

single task/mission duration, total operational time before recharge, mobility, system C2 echelon, level of autonomy, stealth, and physical specifications.

All numerical grades in Appendix E are averages of all responders in each group. All three groups agreed that almost all topic areas were at least "average" (grade of 4), with a preponderance of the grades being 5 (slightly above average) to 7 (expert). Some topics actually had an average of 7.00 from Groups A, B, and/or C, which verifies that all members of the group believed the topic was most important. Given the discussions during the experiment, none of the high grades are surprises. The highest graded topic areas, which support many of the observations in the Scenario Discussions section above, were:

- Ability to Sense: Q5[17]. Friendly/enemy military personnel (Group A). During the experiment, there was much discussion about the need to identify people (friendly and enemy). Also, the typical missions assigned to dismounted infantry platoons are to seek out dismounted personnel. So this high grade was not unexpected.
- Ability to Sense: Q6. Other: humans (Group C). Same as above.
- Information Latency: Q12. Ability to transmit information instantaneously (Group B). Warfighters in the thick of a fight (the members of Group B) want to immediately know when the situation is changing, so it makes sense that they would want information transmitted instantaneously. This same topic was also rated highly by Group A (average of 6.75) and Group C (average of 6.67).
- Single Task/Mission Duration: Q21. Duration up to 48 (mean) hours (Group B). All three groups rated this high, with duration times (Group A: mean of 12.5 hours; Group C: mean of 34 hours) much higher than the highest survey time of "Duration up to 2 hours." A need that kept surfacing during the experiment was using the MAST systems to provide persistent surveillance many hours before the execution of a mission. This would drive the duration time up much beyond 2 hours.
- Mobility: Q32. Ability to keep up with squad movement (Group A). Since the platoon leadership wants to keep abreast of mission status and would not want the MAST systems to slow down the OPTEMPO, it makes sense that they would rate this high. It would also make sense for all members of Group B to want the MAST systems to keep up with them. Group B did provide this capability area a high score (average of 6.00 for this question) and it also ranks high in Section B (discussed below). Perhaps it was not a perfect score because some squad leaders expressed the desire for the MAST systems to "stay out of their way" once the fighting started.
- Stealth: Q50. Other type of stealth (write in of "camouflage") (Groups B and C). Group A was also high, with an average of 6.67. In terms of camouflage, the groups wanted the MAST systems to blend in with the local area, perhaps looking like a piece of a building, a local insect, or a piece of typical trash.

Of particular interest are the few topics that had notably diverse grades. In particular, for Q16 (Ability to return with downloadable information), Group A had an average grade of 6.50, while Groups B and C had averages of 3.00 and 3.56 respectively. The platoon leadership would want

---

[17] The term "Q5" does not refer to Question #5, but rather to item #5 in Section A. In this case, Q5 is under the grouping "Ability to Sense": and it is "Friendly/Enemy military personnel." This is similar for all other references to Q# - e.g., Q31 is under the grouping "Mobility" and it is "Ability to keep up with squad movement."

downloadable detailed information (to include images) for a few reasons: 1) to obtain early detailed information to support mission planning, 2) to support "after action reviews" of a mission, and 3) to use the mission to populate operations and intelligence databases and tools (i.e., the Tactical Ground Reporting (TIGR)[18] system).

Another notable difference was in the mobility area. For all of the questions, Groups A and B scored the importance high, but for three questions (Q31, Ability to do all mobility maneuvers listed above; Q32, Ability to keep up with the squad movement; and Q33, Ability to keep up ahead of squad), Group C was somewhat lower. Perhaps this difference can be attributed to the wishful desires of the military personnel as opposed to the "reality" of what the MAST personnel know can be accomplished.

One final question (Q44, System is tele-operated and all decisions are made by the operator) was rated relatively high by Group A (average of 5.88) while Groups B and C had averages of 4.57 and 3.63, respectively. This may be attributed to the need of the platoon leadership to want to maintain control of the MAST systems while Group B doesn't want to be bothered with them once the fighting starts and Group C desires to achieve the technical challenge of autonomous actions.

Section B. Since all capabilities were ranked relatively high in Section A, it is difficult to use the information to prioritize the efforts of the MAST program. Section B, on the other hand, provided more clarity since it asked the three groups to rank order the ten capability areas from 1 to 10, with no ties.

For Group A, the platoon leadership gave high rankings (average of 7.00 and above) to information fidelity (7.71), mobility (7.00), and stealth (7.00), with mobility having the smallest standard deviation (1.83) of all ten areas. These three areas are supportive of the platoon leadership's desires. First, information fidelity supports the need to acquire actionable information for mission planning and execution. Mobility is important to get the MAST systems in the right place at the right time and to not slow down the execution of the operation. Finally, stealth is important to ensure that the enemy is not alerted to an impending action or the direction of an executed action. Group A also gave low scores (average of 3.00 or lower) to single task/mission duration (3.00) and physical specification (2.71). The platoon leadership must plan for the execution of multiple tasks within a single mission, as well as the possibility of immediately transitioning to newly assigned, follow-on missions. Perhaps the area of physical specification was rated low because it is not focused on a tactical capability like the other nine capability areas.

For Group B, high marks were given to the same three areas as Group A, but with notably higher averages: information fidelity (8.71), stealth (7.57), and mobility (7.50), with the lowest standard deviation of 1.11 aligned with information fidelity. Although stealth was slightly higher than mobility, they were virtually equal, as in Group A. Group B is closer to the fight. Its members

---

[18] TIGR is a multimedia reporting system for Warfighters at the patrol level, allowing users to collect and share information to improve situational awareness and to facilitate collaboration and information analysis among junior officers and non-commissioned officers. It is similar to popular social networking tools found on the Internet today, but it operates on a classified network.

are the Warfighters who would have to move to the objective and kick down the doors to rooms. To them, actionable information means knowing, before they enter a room, who might have a weapon and where those people may be in the room, so information fidelity is very important to them. Squad members are also more concerned about the MAST systems keeping up with them and not giving away their intentions. Group B gave low scores to system C2 echelon (2.83) and physical specification (2.83). A few of Group B's members stated that the C2 and maintenance of the MAST systems should be at the platoon level, with a platoon robotics NCO. This may have influenced this score. Additionally, C2 is inherently more of a concern at the platoon and squad levels than at the fire team level, so perhaps that has some influence as well. As with Group A, the low score of physical specification may be likewise attributed to the fact that it is not a tactical capability area.

Group C's ratings were a little different than Groups A and B. Group C's highest scores (highest to lowest) went to stealth (8.89), ability to sense (8.56), and mobility (7.88). On average, these scores are much higher than Groups A and B. Also, for all ten capability areas, Group C had relatively smaller standard deviations, with eight less than 2.00. In addition to the high marks for stealth, ability to sense, and mobility being attributed to Warfighter discussions over the two days, they might also be attributed to their associated technical challenges. Low scores were given to system C2 echelon (2.25) and physical specifications (1.88). These scores are probably low for similar reasons described in Group B.

Since MAST systems will depend heavily on "level of autonomy," it is interesting that it received slightly less than average scores from Group A (4.71) and Group B (4.83), and not the highest score from Group C (6.83). It is not certain why Groups A and B were so low. Perhaps it was assumed that the other critical high-score capability areas would readily have autonomy. Similarly, with respect to physical specifications, autonomy does not sound like a warfighting capability area. Finally, it may have been due to a lack of understanding of or appreciation for autonomous behavior versus semi-autonomous behavior or even tele-operation. For Group C, its members were probably swayed by the discussions of the Warfighters – which is a tribute to the Gedanken Experiment process.

Sections C through G. The comments provided by the participants to the topics of Sections C through G are quite varied and interesting. They are listed in detail in Appendix E for review. Some interesting points are:

- Section C. It appears to be difficult to assess TTP needs within a Gedanken Experiment, where system capabilities are merely speculations. TTP development should keep pace with the technical development of the MAST systems. Some worthwhile feedback was provided with respect to C2, but the Warfighters gave little information that would assist in system design. On the other hand, it appears that Group C developed an appreciation for TTP in the tutorials, hands-on training, and the experiment that will greatly benefit them in their development of the MAST systems.
- Section D. There was significant feedback from all three groups on what is liked about the MAST system concept. Among the groups, recurring feedback included sensing (including early warning and local security), mobility, size, and keeping Warfighters out of harm's way.

- Section E. There wasn't as much feedback from Groups A and B on recommended improvements. Ideas included enhancing sensors (including audible sensing on systems and controllers), adding lethal capabilities, and being tamper-proof. Group C had significant feedback, which may again be a tribute to experiment.
- Section F. For topics that were not discussed, Groups A and B feedback included IEDs, identification between friend and foe, lethality, and don't consider replacing Warfighters with MAST systems. As in the topics above, Group C had a significant amount of feedback.
- Section G. For additional comments, the Warfighters saw value in the effort and wanted the capabilities much sooner than the 20 years discussed for this particular effort. They also appreciated participating in the experiment. Group C saw a need for spiral development of the system, to get capabilities out to the Warfighters at a faster pace.

In summary, the survey provided some useful information. Priority of effort appears to be information fidelity (and sensing), mobility, and stealth (perhaps as simple as camouflage). To achieve the tasks inherent with the feedback, there is obviously an implied need for autonomous, or at least semi-autonomous, behavior. Additionally, the Gedanken Experiment appeared to be of benefit to all participants, especially the MAST personnel in Group C.

End-of-the-Day Discussions

At the end of each day, the floor was open to general discussions. Most of the discussions addressed topics already identified above. However, a few additional topics were raised. Two topics addressed policy and political or cultural advantages. Operational time was a third topic.

From a policy perspective, it was felt that directives and regulations needed to be considered to control the misuse and abuse of MAST systems. For example, MAST systems should not be used during free time in robot gladiator competitions. Additionally, they should not be used to invade an individual's privacy or to conduct outright theft.

From a political or cultural advantage, the question was raised as to how the MAST systems could contribute to winning hearts and minds of the local civilians. Intuitively, the capabilities offered by MAST platforms could support enhancements to the irregular warfare (IW) line of operation (LOO) of security. They could reduce U.S. force offensive exchanges, thus minimizing casualties in the native civilian populace. This advantage could be enhanced by using MAST systems to improve "surgical strike" outcomes, minimize false alarms, and better identify friend or foe. There were no recommendations as to how MAST systems could be used in support of other IW LOOs (i.e., governance, rule of law, and economic development).

As far as operational time was concerned, the participants equated this to battery life. Short-term battery life is adequate for raid functions – door breach, deception, short-term surveillance. Long-term battery life would be really useful for persistent surveillance, and 24 to 48 hours of battery life would be ideal in a lot of situations.

Summary and Noted Benefits

The overall effort was a successful endeavor for the MAST CTA. The MAST CTA personnel gained much from their exposure to small unit urban operations, including the first day's classroom tutorial and the hands-on training experience at an urban dismounted infantry training site. Their involvement in the Gedanken Experiments on the second and third day:

- Exposed them to Warfighter operational thought processes and concerns
- Exposed them to Warfighter TTP, which generated ideas on how MAST systems might be integrated into small unit urban operations
- Utilized their subject matter expertise to answer the technical/capability questions of the Warfighter participants

In addition, the Warfighter participants gained insights on what the Army's science and technology community is trying to do for them. They greatly appreciated the opportunity to participate in the exercise. It was interesting to note that the younger Warfighter participants were less inhibited by doctrine and more willing to be imaginative in terms of capabilities and use.

Many technical and operational topics are highlighted in this paper, and can't all be summarized here. The following were of most benefit to the MAST CTA CAM:

- The Warfighters' highest capability priorities were information fidelity (including sensing), mobility, and stealth. In this case, stealth is more of a camouflage or blending in with the surroundings and reducing noise, as opposed to reductions in visual, electromagnetic, and radar signatures.

- For most areas, the platoon level leaders and the squad level leaders agreed on uses and needs. There were some notable differences, especially in the area of information fidelity (platoon level leaders wanted more all the time), real-time feedback (platoon level leaders were willing to trade fidelity for real-time feedback), C2 of the systems (during mission execution, squad level leaders wanted hands off, while platoon level leaders wanted squads to maintain operational control), and overall control (squad level leaders wanted a hierarchy of control and maintenance at platoon level, while platoon level leaders saw all control and maintenance at the squad level).

- For TTP, novel ideas for deployment were discussed and concerns for C2 kept arising. With respect to C2, there is a need for an easy-to-use Warfighter-MAST systems interface that supports Warfighter-machine interactions, enhances Warfighter situational awareness, and allows for changes in operational control during the execution of a mission.

- Additionally for C2, there is a need for scalable interfaces that provide differing views of C2 for platoon level leaders and squad level leaders, or the differing needs for levels of information fidelity during an operation. For example, platforms can best be exploited for intelligence gathering to aid in planning but use should be limited during actual combat. To accommodate these differences, multiple MAST designs may be needed along with scalable levels of autonomy among an ensemble of MAST systems of varying types, so that autonomy levels can be increased or decreased based upon the mission phase. For example, when the MAST technology is immature and autonomous behavior is not very predictable, it may be best to scale down autonomy during a firefight.

- Although lethal capabilities were discussed, there appeared to be more of an interest in non-lethal capabilities to support deception and diversion during offensive operations and to possibly reduce the number of civilian casualties, especially in hostage situations. For the MAST CTA CAM, this was an unforeseen beneficial warfighting capability that is potentially doable for micro autonomous systems.

The most important benefit for the MAST CTA CAM was his use of the Warfighters' perspectives to assess the benefits of design approaches. For example, platforms can be best exploited for intelligence gathering to aid in planning but should not be used when the potential for combat operations is high. For cultural reasons (i.e., military culture) and the immaturity of the technology, Warfighters should not be teamed with autonomous platforms during a firefight. This single perspective has helped MAST CTA CAM keep the researchers focused on a well defined design approach.

The MAST CTA CAM acknowledged that not only did the Gedanken Experiment output influence his leadership of MAST; but as ARL defines its strategic vision in autonomous platforms of all sizes and shapes, it has also provided him supporting information to debate with other researchers in the lab about the reasonableness of promoting room clearing by autonomous platforms. He believes he has successfully argued that this is not a good application to consider, at least not until the technology has matured and Warfighters trust the capabilities of such platforms.

The overall experience was very beneficial to all participants in the exercise. It is highly recommended that the MAST CTA keep a close working relationship with Warfighters, especially those involved with developing TTP and assessing Warfighter-machine interfaces at Fort Benning, GA.

Appendix A. Offensive Scenario

Urban Operations with Small Combat Units Equipped with Micro Autonomous Robotic/Autonomous Systems

Situation

A Small Combat Unit (SCU) has been given the task to clear two buildings in a small urban area, known to be meeting locations for local insurgent forces.

Intelligence Information

Physical Environment

The setting for this scenario is in the northeast section of a small urban town during the hours of darkness. The roads are paved and unpaved. The buildings vary in height and shape. Primarily, the buildings consist of one or two stories, and there are multiple rooms per floor. A layout of the mission area is depicted in Figure 5.

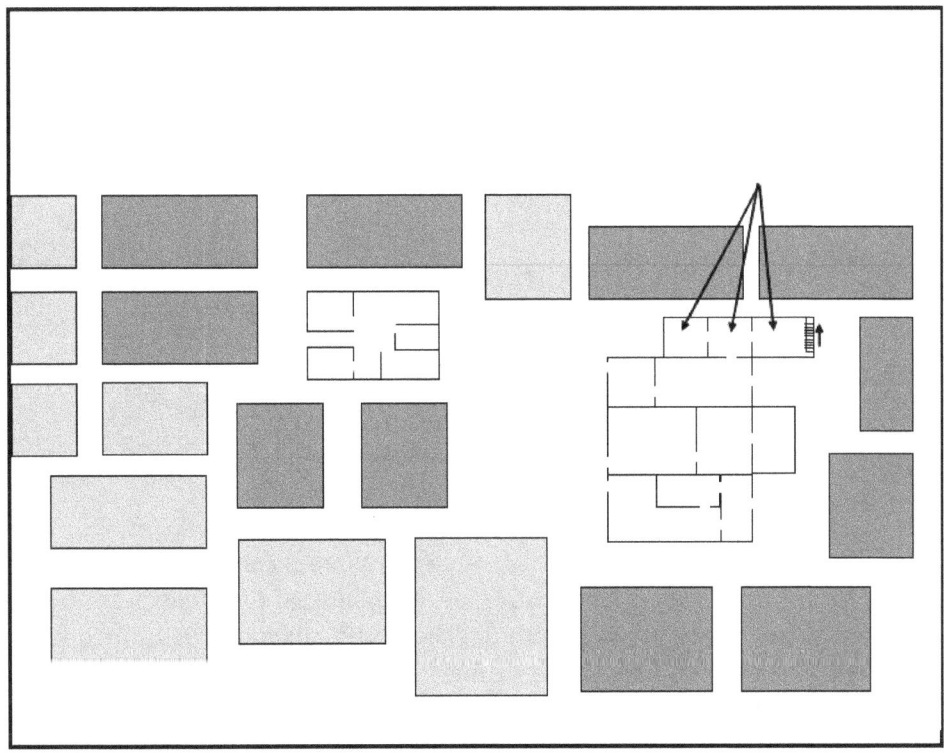

Figure 5. Physical Environment

Building A is believed to be a single residence with about 5 rooms. Building B is a multi-family building that has three separate residences. The number of rooms per residence is not known. The northern residence (R1) has access to its roof.

Enemy and Civilian Situation

Current intelligence reports identify Building A and Building B's Residences R1 and R2 being occupied with both insurgents and neutral civilians. Building B's R3 is occupied with at least neutral civilians.

The leader of the insurgent forces is suspected to be located in one of the buildings. The exact number of insurgents located in each building is unknown; however, it is suspected that there could be two (2) to ten (10) insurgents within Building A and five (5) to fifteen (15) insurgents in Building B.

A depiction of a possible insurgent and civilian situation in Buildings A & B is in Figure 6.

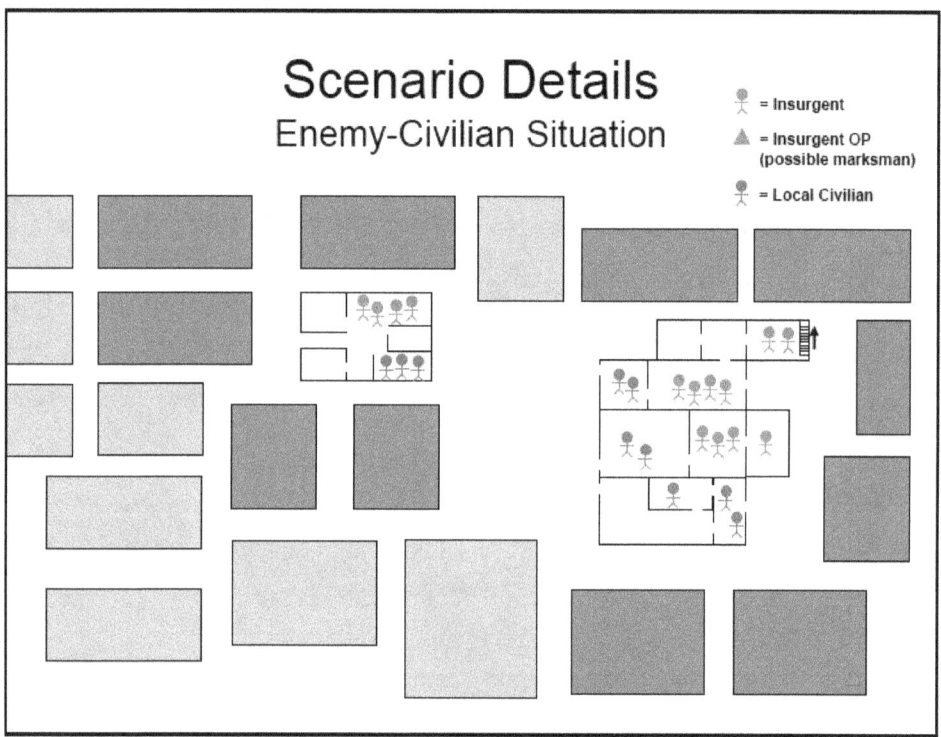

Figure 6. Possible Insurgent and Civilian Situation

Aerial intelligence assets have also identified as many as three insurgent OPs on the roofs of tall buildings. The latest information on the location of the insurgent OPs was used to plot the locations (identified as red triangles) in Figure 6. These OPs may also be trained marksmen. It is unlikely that the OPs will open fire. Their primary objective is to alert the insurgent leader of the presence of U.S. forces. However, it is possible that the insurgent leader may order them to open fire.

Since the operation takes place at night, it is possible that the insurgents could be unaware of the platoon's approach; however; it should be assumed that the OPs will alert the insurgent leader. The insurgents have very few night vision devices, and they are probably assigned to the OPs.

Human Intelligence (HUMINT) reports that the insurgents are here for a meeting of multiple insurgent cells. Therefore, it is possible that other cell personnel might be arriving during the course of the night.

Once the insurgents know that Blue Forces are patrolling in the vicinity of their meeting locations, their most likely course of action is to establish defensive positions within the buildings to protect their leader. It is also possible that some of the insurgents will defy their leader's orders and run; therefore, it is important for the platoon to observe the rear of the buildings to prevent insurgents from running.

The local civilian population in this portion of the town fears the insurgent leader and typically remains indoors during hours of darkness. While there is not an enforced curfew, the local populace knows that the insurgents typically conduct meetings and emplace IEDs during the hours of darkness. Therefore, they have made it a practice to remain indoors at night. It is possible for neutral civilians to be outdoors. Most likely civilians are in Buildings A and B. It is important that Blue Forces properly identify their targets prior to engaging.

Blue Forces

An SCU, which is a conceptual dismounted infantry platoon for this mission, is organized as depicted in Table 1.

Headquarters Section	1st Squad	2nd Squad	3rd Squad	Weapons Squad	Robotics Section
Plt Ldr	Sqd Ldr	Sqd Ldr	Sqd Ldr	Sqd Ldr	Robotics NCO
Plt Sgt	Fire Tm Ldr A	Fire Tm Ldr A	Fire Tm Ldr A	Javelin Gnr	SUGV1
RATELO	M249 Gnr	M249 Gnr	M249 Gnr	Asst Javelin Gnr	SUGV2
Plt Medic	Grenadier	Grenadier	Grenadier	M240 MG Gnr	SUGV3
	Rifleman	Rifleman	Rifleman	Asst M240 MG Gnr	
	Fire Tm Ldr B	Fire Tm Ldr B	Fire Tm Ldr B	Javelin Gnr	
	M249 Gnr	M249 Gnr	M249 Gnr	Asst Javelin Gnr	
	Grenadier	Grenadier	Grenadier	M240 MG Gnr	
	Rifleman	Rifleman	Rifleman	Asst M240 MG Gnr	

Table 1. Small Combat Unit

Note 1: Asst = Assistant   Plt = Platoon   RATELO = Radio-Telephone
Gnr = Gunner   Sgt = Sergeant   Operator
Ldr = Leader   Sqd = Squad   SUGV = Small Unmanned
MG = Machine Gun   Tm = Team   Ground Vehicle

Note 2. There is no current doctrine or published organizational structure for the inclusion of micro unmanned and autonomous systems (UAS) in an SCU. It is assumed that the participants in the Gedanken Experiment will provide insights on the organizational structure and use of micro UAS. A possible structure for assigning micro UAS is depicted in Table 2 – with the micro UAS identified as the ARL MAST systems (in red text).

Note 3: This assignment of MAST systems is just a simple recommendation made by the author of the scenario. It is unknown if squad leaders would realistically have MAST systems, since their duties and responsibilities will most likely be focused on commanding and controlling the squad and not on maneuvering MAST systems and maintaining them. The squad leader may be directing the use of the MAST systems. As an analogy, squad leaders carry rifles and not equipment with higher fire power (i.e., machine guns (MGs) and grenade launchers). The participants of the Gedanken Experiment will be solicited for recommendations on the assignment of MAST systems.

Headquarters Section	1st Squad	2nd Squad	3rd Squad	Weapons Squad	Robotics Section
Plt Ldr	Sqd Ldr w/MAST(s)	Sqd Ldr w/MAST(s)	Sqd Ldr w/MAST(s)	Sqd Ldr w/MAST(s)	Robotics NCO
Plt Sgt	Fire Tm Ldr A	Fire Tm Ldr A	Fire Tm Ldr A	Javelin Gnr	SUGV1 w/MAST(s)
RATELO	M249 Gnr	M249 Gnr	M249 Gnr	Asst Javelin Gnr	SUGV2 w/MAST(s)
Plt Medic	Grenadier	Grenadier	Grenadier	M240 MG Gnr	SUGV3 w/MAST(s)
	Rifleman w/MAST(s)	Rifleman w/MAST(s)	Rifleman w/MAST(s)	Asst M240 MG Gnr	
	Fire Tm Ldr B	Fire Tm Ldr B	Fire Tm Ldr B	Javelin Gnr	
	M249 Gnr	M249 Gnr	M249 Gnr	Asst Javelin Gnr	
	Grenadier	Grenadier	Grenadier	M240 MG Gnr	
	Rifleman w/MAST(s)	Rifleman w/MAST(s)	Rifleman w/MAST(s)	Asst M240 MG Gnr	

Table 2. Potential Small Combat Unit Organization with MAST systems

Mission

The SCU is to clear Buildings A and B, known meeting locations for the local insurgent forces. The SCU will
- Conduct an assault from the southwest corner of the town to OBJs Homer (building A) and Bart (building B)
- Establish SBF positions covering the entrances to the buildings and prevent a counterattack by insurgents from OBJ Bart (building B) while OBJ Homer is being taken
- Establish OPs covering the rear exits of OBJ Bart (building B)

- Conduct initial reconnaissance of buildings/rooms using SUGVs and MAST systems
- Neutralize insurgents

The primary task during this clearing operation is to identify and neutralize (capture or kill) all insurgents located within the objectives, while minimizing the number of civilian casualties.

Initial Blue Positions and Control Graphics

The initial location of the Blue Forces and the Control Graphics are depicted in Figure 7. The initial location of the Weapons Squad is not depicted.

1st Squad and 2nd Squad will be assaulting OBJ Homer. 3rd Squad will be kept in reserve. On order, once Homer is taken, 3rd Squad will initiate its assault on OBJ Bart.

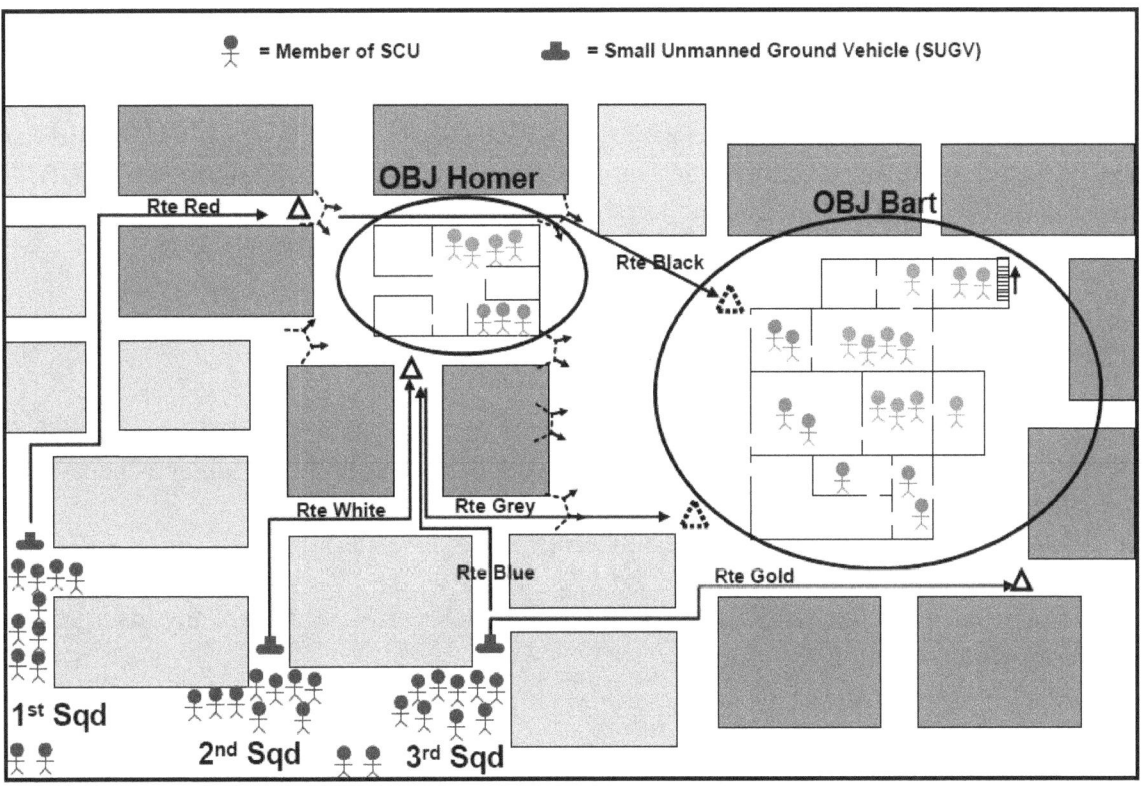

Figure 7. Initial Blue Locations and Control Graphics

Phase I – OBJ Homer

Step 1: Initiation of the Assault
Initially, the Robotics NCO will operate all three SUGVs moving them down three separate routes to set up OPs. All three SUGVs will move at the same time.
- SUGV1 (with 1st Squad) will move north along Route Red to set up an OP to identify insurgent activity moving in from the northwest or exiting OBJ Homer.
- SUGV2 (with 2nd Squad) will move along Route White to set up an OP to identify insurgent activity on the south side of or exiting from OBJ Homer.

– SUGV3 (with 3rd Squad) will move along Route Gold to set up an OP to identify any insurgent activity moving in from the east or exiting OBJ Bart.

Specific Question:
How should the Robotics NCO use the MAST systems?

Step 2: Once the SUGVs are in place and the area is clear, 1st Squad will move along Route Red and 2nd Squad will move along Route White. 3rd Squad will initially remain in reserve.

Step 3: Once 1st and 2nd Squads are in location, SUGVs 1 and 2 will move toward OBJ Homer to begin a reconnaissance of the exterior of the building.
Specific Questions:
Are MAST systems used here?
If so, how?
What type: ground and/or air?
What about the insurgent OPs?

Step 4: Once 1st and 2nd Squads are in position, the Weapons Squad will set up a MG and Javelin in an SBF position (depicted as small, dashed-line figure with two arrows) northwest of OBJ Homer and a MG and Javelin in the southwest of OBJ Homer. Based on the activity in OBJ Bart, the Weapons Squad will be prepared to move east to alternate SBF positions.
Specific Questions:
Does the Weapons Squad need MAST systems?
If so, how will they be used?

Step 5: Once the exterior of OBJ Homer is clear, 2nd Squad will move to OBJ Homer and begin to clear. 1st Squad will provide local security and assist with clearing OBJ Homer as necessary.
Specific Questions:
What should be the priority of use of MAST systems?

Assume the SUGVs and the MAST systems are used to determine if there are any insurgents or civilians located in the building.
Specific Questions:
If humans are detected, what actions are taken?
What are the TTP?
Must the humans be identified as insurgents or neutrals?
If insurgents are identified, how should the information be used?

Step 6: Complete the clearing of OBJ Homer.

Phase 2 – OBJ Bart

Once OBJ Homer is clear:
– SUGV1 will move along Route Black to set up a secondary OP
– SUGV2 will move along Route Grey to set up a secondary OP

- 1st Squad will move half way down Route Black to the vicinity of the SBF position
- 2nd Squad will move down Route Grey to vicinity of the SBF position
- 3rd Squad will move along Route Gold to prepare to secure the southern and eastern ends of OBJ Bart
- The Weapons Squad will move due east to set up 4 separate SBF positions

Once all are in position, 3rd Squad will move forward to begin clearing the southernmost residence of OBJ Bart. Once that is clear, it will move to clear the next residence. Finally the northernmost residence will be cleared by a combination of 2nd and 3rd Squads.

NOTE 1: As done in OBJ Homer, walk through the steps of this clearing operation, asking questions similar to those identified in clearing OBJ Homer.

NOTE 2: The clearing of the roof needs to be discussed.

Once OBJ Bart is cleared, this scenario is completed.

Scenario Excursions:
1. Insurgents attempt to escape from OBJs Homer and/or Bart
2. SUGVs 1 and 2 break down or are neutralized by the insurgents and only MAST systems are working
3. Windy conditions outside, so flying MAST systems may be difficult to control – if they can fly at all.

Appendix B. Defensive Scenario

Urban Operations with Small Combat Units Equipped with Micro Autonomous Systems

Situation

The SCU conducts the offensive operation discussed in Appendix A. Once
OBJ Homer is cleared, the platoon receives a report that an additional ten (10) to twenty (20) insurgents are moving into the area. The company commander orders the platoon to take up defensive positions and await further orders. The rest of the company is moving to the platoon area of operation to conduct company level operations.

Mission

The platoon establishes a defensive position
Specific Questions:
How will the MAST systems be used to support this specific task?
If fighting ensues, how are the MAST systems used?

Appendix C. Stabilization Operation Scenario

Summary of the Scenario

An IED explosion occurs in a large city and snipers are shooting at emergency responders. A reaction force (a U.S. infantry company augmented with a platoon made up of local nationals) is sent to secure the area and neutralize the snipers. Along the way the reaction force has to identify and deal with obstacles (crowded market, protestors, etc.). Once at the IED scene, the reaction force discovers that in addition to the snipers, local civilians may have been abducted and are being detained by insurgents. The reaction force receives additional orders to work with local police forces to verify the abduction, then locate and free the civilians.

Background Discussion

Some of the unique characteristics of stabilization operations are:
- Rules of Engagement (ROEs) are far more restrictive than in major combat operations so as to minimize "media opportunities" and prevent collateral damage. This severely constrains the use of large caliber munitions and perhaps armed robotic systems. There may be an impact on interactions of robotic systems with humans. Movement of robotic systems into homes/rooms may be considered an infringement on privacy.
- The Warfighter is more a "dismounted policeman" than a mounted or dismounted warrior. Within a city, tanks and armored fighting vehicles are often parked in motor pools and crews and squads are assigned to dismounted patrols.
- Maneuver operations are highly restricted since armored vehicles may damage streets, homes, automobiles, etc. Additionally, streets are often cluttered with day-to-day activities that should not be interrupted with military operations. Finally, units are assigned to static bases.
- There is much more background clutter in the form of radio transmissions, lights, pedestrians, civilian automobiles, and other interferences that degrade the performance of military communications, sensors, and human sensing. For example, numerous lights at night interfere with the performance of night vision goggles. Electromagnetic emissions may interfere with intra-/inter-robotic communications.
- The air space is very complex, due to the much greater use of manned and unmanned military and civilian aircraft, as well as the occasional use of non-line-of-sight munitions. It is very difficult to deconflict the airspace, which often deters the use of unmanned aerial systems by battalion and below forces.
- The probability of Warfighters interacting with civilians is high so there is a much greater requirement for Warfighters to understand the local populace in terms of language and culture.
- There is a much greater concern for coordinating diplomatic, information operations, military, and economic (DIME) actions and their effects on political, military, economic, social, information, and infrastructure (PMESII) systems.

Stabilization Operations with Small Combat Units Equipped with Micro Autonomous Systems

General Situation

The United States has recently completed combat operations in a foreign country and has transitioned to stabilization and reconstruction operations. Formation of a new government is underway, as well as the development of its military and police forces. U.S. stabilization forces have been positioned throughout the country in static, battalion-size bases and have been using peacekeeping, cooperative activities, and coercive actions to maintain order in cities and provide security for U.S.-sponsored reconstruction operations in their assigned sectors. U.S. funded reconstruction operations are primarily focused on repairing or upgrading the utilities and transportation infrastructure. All stabilization missions are coordinated and conducted with the participation of the local national military and police forces, as well as non-Department of Defense (non-DoD) U.S. entities.

Insurgent activities continue to plague U.S. and local government efforts to rebuild the country. Current intelligence indicates there are three major insurgent factions:

- Non-state actor terrorist organizations comprised of a combination of local nationals and foreign insurgents who have infiltrated the country in question.
- Ethnic/religious militant groups. These ethnic/religious factions are closely related in terms of ancestry and beliefs. However, they believe that a few sharp differences exist that justify violence toward each other. Local national police and military forces are often established with a predominance of one ethnic/religious group or another.
- Criminal groups. These groups attempt to use violence to ensure money is provided to them in the form of contracts, grants, management fees, and even illegal payments.

All three insurgent factions use a combination of IEDs, snipers, and small paramilitary forces to promote violence.

Specific Situation. An IED has exploded in a market place. Local emergency responders have arrived; but triage, medical, utility restoration, and clean-up efforts are being hindered by at least two snipers. A U.S. reaction force (an infantry company, about 160 personnel), augmented with a local national infantry platoon (about 40 personnel) and U.S. medical personnel, is tasked to move to the scene to help restore order and assist with medical needs. The infantry company commander and platoon leaders are to make contact with the local police force upon arrival and coordinate efforts with them. The attack occurred mid-afternoon when the market place is extremely crowded with pedestrians, automobiles, and street vendors. The market place has one- to five-story commercial stores on either side of the street and is frequented primarily by one of the warring religious groups. This may be an issue since the augmented infantry platoon is representative of the opposing religious group. Additionally, there have been reports of civilians gathering in large groups in nearby neighborhoods to protest the attack. Historically, these gatherings become violent when they come in contact with opposing religious groups. Soon after arriving at the scene, the company commander receives unconfirmed reports from locals that an additional complication may have arisen. It has been reported that a local official traveled to the scene to view the area, and while there, he and his driver were abducted by the insurgents. It is suspected he and his driver are being kept against their will somewhere in a block of residential housing adjacent to the scene of the IED attack. The local neighborhood buildings surrounding the market place are comprised mostly of one to two story residential homes closely packed together. After reporting this information to the battalion commander, the company commander is ordered to work with local police to resolve this new problem. Efforts are prioritized in the following order: locate and neutralize the snipers; secure and support humanitarian assistance at the IED incident; verify the reported abduction; if the abduction occurred, find and assist local forces in recovering the official and his driver. Some of these actions can occur simultaneously. Note that this is the Three Block War[19] scenario.

Mission Steps: To accomplish this mission the augmented U.S. infantry company must accomplish the following activities:

Step 1: Plan operations. Static (prior to movement) as well as dynamic (en-route) planning and replanning. Assessment of the mission, enemy, terrain, troops, and time available (METT-T). Integration of available military assets, non-U.S. assets, unmanned systems (robots and sensors), latest terrain/feature data, and PMESII systems. Battlespace awareness, course-of-action analysis, mission selection, and mission rehearsal.

Step 2: Move to IED scene. While en-route, protestors begin to gather in a road that is part of the planned route for the reaction force. The situation at the IED site changes—insurgents abduct a local official and his driver. Abduction is seen by locals, but because of the confusion with dealing with casualties, fires, and building damage, this information is not reported to proper officials.

Step 3: Secure the area. The reaction force arrives on the scene. Perimeter security is established using a combination of vehicles, dismounted personnel, sensors, and unmanned ground and air systems. Ad-hoc traffic control points are established. Civilians' identities are checked against databases.

Step 4: Locate, identify, and neutralize the snipers. Using a combination of HUMINT, signal intelligence (SIGINT), imagery intelligence (IMINT), urban terrain analysis systems, and historical data, identify likely and known sniper locations. Sniper location systems are used to locate the sniper when shots are fired on the reaction force. Once identified, non-linear tactics and networked weapons systems are used to neutralize the sniper. While the reaction force is accomplishing this task, a second sniper opens fire and the reaction team adjusts to neutralize this additional sniper.

Step 5: Assist with medical needs. While military actions are taking place against the snipers, assistance is being provided at the scene of the IED attack. Military personnel assist local emergency responders with their duties. Because of the devastation, additional U.S. emergency responders are brought to the area. The local commander must be able to coordinate

---

[19] The Three Block War was conceived by General Charles Krulak to describe how the full spectrum of war could be faced simultaneously by Warfighters in lower echelon units. In three adjacent city blocks, Warfighters may be required to conduct full-scale military action, peacekeeping operations, and humanitarian relief.

humanitarian assistance being provided by his personnel, the newly arrived U.S. personnel, and the local emergency responders while neutralizing the snipers and setting up perimeter security. Additionally, in these situations, U.S. personnel have to consider personal medical safety, since in addition to the dangers of military attacks, there may be situations where local civilians may be afflicted with a contagious disease.

Step 6: Verify, locate, and assist with freeing the detained civilians. Once the sniper situation is under control, the company commander has to respond to a new order to verify, locate, and assist with freeing the detained local official and his driver. The activity within this scenario repeats many of the activities mentioned above—plan the operation, move to the area of operations, secure the area, and then assist with the detained civilians. The major difference from the "neutralize the sniper" activity is that here the local security forces lead and execute the effort with U.S. military forces in a supporting role and subordinate to the C2 of the local security force leader, who is a policeman. It is hoped that military actions will be kept to a minimum. If military actions are needed, the military commander should minimize collateral damage by using actionable information such as the location of the building and room holding the detained civilians and their abductors, as well as their approximate location and physical orientation within the room.

Step 7: Treat and evacuate casualties. Medical attention is provided to U.S. casualties as they occurred during the mission. Triage and life-saving actions were taken, as needed. Casualties needing evacuation are moved back to aid stations as rapidly as possible. Assessment of air evacuation must take into account urban structures which may be too high and close together for helicopters to get near the company's area of operation.

Step 8: Consolidate forces. After operations are completed and the situation is fully under control by local police and medical personnel, the company commander is ordered to consolidate his forces. This requires rapid assessment of the location and activities of all personnel and equipment (i.e., unmanned systems) involved with humanitarian assistance, perimeter security, sniper neutralization, and freeing the detained civilians. All assets (i.e., sensors and unmanned systems) must be retrieved.

Step 9: Depart the area. In planning an evacuation route, the company commander must work closely with battalion headquarters to assess routes in the same fashion as the initial planning and movement tasks above.

Note: Through each of the above steps, participants are asked questions similar to those in the offensive scenario.

Appendix D. Demographics of Participants

GROUP/SAMPLE SIZE:

A – Platoon Sergeant (PLT SGT) / Platoon Leader (PLT LDR)   = 8
B – Team Leader (TM LDR) / Squad Leader (SQD LDR)   = 7
C – Civilian   = 9

	A	B	C
MOS			NA
- 11A (Infantry Officer)	4	-	
- 11B (Infantryman)	3	6	
- 19D (Cavalry Scout)	1	1	
RANK	-	-	NA
- Captain (CPT)*	4	-	
- Sergeant (SGT)	-	1	
- Staff Sergeant (SSG)	-	4	
- Sergeant First Class (SFC)	4	2	
* PLT LDR as a Lieutenant			
DUTY POSITION			NA
The following were of most benefit to the MAST CTA CAM:Drill SGT	-	5	
–S3 (Operations)	1	-	
–NCOIC	1	1	
–PLT SGT	1	-	
–PSG/GNR	1	-	
–Ranger Instructor	1	-	
–Student	1	-	
–TM LDR	-	1	
–No Response	2	-	

1. How long have you served in the military? (Months)

MEAN RESPONSE		
A	B	C
133	88	NA

2. What was the last military professional development course you completed?

	A	B	C
AIT	-	-	NA
PLDC	-	-	NA
BNCOC	-	5	NA
ANCOC	4	1	NA
IOBC	2	-	NA
IOAC	-	-	NA
ICCC	-	-	NA
Other:	-	-	NA
– IMLC	1	-	NA
– ITAS/ABIC	1	-	NA
– DRILL SGT SCH	-	1	NA

3. Do you have combat or stabilization operations experience? If yes, what was the location, duty position, and length of stay?

	A	B	C
Yes	8	7	1
No	0	0	5
NR	0	0	3

Comments   **No. of Responses**

A

Afghanistan – platoon sergeant – 4 months	1
Baghdad – platoon sergeant, 50 cal machine gunner – 16 months. On the division MTT team (gunner)	1
Baghdad – MTT XO – 9 months	1
Baghdad – battle captain – 15 months	1
Baghdad – platoon leader – 9 months	1
Baghdad – battalion S1 – 12 months	1
Bosnia – squad leader – 8 months	1
Iraq – section commander and intelligence (S2) advisor – 12 months	1
Iraq – platoon leader (mechanized infantry) – 7 months	1
Iraq – weapon squad leader – 15 months	1
Iraq – 12 months	1
Iraq – 5 months	1
Iraq – 13 months	1
Kosovo – 7 months	1
Iraq – 13 months	1
Transition team advisor – 12 months	1
Assistant team leader on a national police transition team.	1

Comments	No. of Responses

B

Afghanistan - .50 cal gunner, team leaders – 6 months	1
Afghanistan – scout squad leader – 16 months	
Iraq – driver/team leaders – 12 months	1
Iraq, Afghanistan, Pakistan – team leader – 2 years	1
Iraq – squad leader – 6 months	1
Iraq – squad leader – 13 months	1
Iraq – squad leader – 12 months	1
Iraq – team leader – 12 months	1
Iraq – rifle team leader – 12 months	1
OIF – MTT advisor, team leader	1

C
NA

4. Do you have experience using robotic vehicles in military operations?

	A	B	C
No	5	6	4
Yes	3	1	2
NR	0	0	3

A
MARCBOTS in IED reduction ops.   1

B
Robots that detect IEDs.   1

5. Using the scale below, please rate your level of experience with the following computer software and computer related activities.

1	2	3	4	5	6	7
No experience	Below average	Slightly below average	Average	Slightly above average	Above average	Expert

	MEAN RESPONSE		
	A	B	C
a. Microsoft Windows 98, 2000, XP, etc.	3.75	4.79	6.71
b. Computer based games	3.87	4.00	5.14
c. Army digital systems (e.g. FBCB2[20])	4.13	4.21	1.33
d. I would self rate my computer skills as:	3.75	4.21	6.43

---

[20] Force XXI Battle Command Brigade and Below (FBCB2) is a C2 platform that tracks the location of forces.

6. Self-rating of Knowledge, Skills, and Abilities (KSA) related to Infantry duties. Please use the scale below to rate the following skills.

1	2	3	4	5	6	7
No experience	Below average	Slightly below average	Average	Slightly above average	Above average	Expert

	MEAN RESPONSE		
	A	B	C
a. Knowledge of tactics, techniques, and procedures (TTP).	5.88	5.79	1.71
b. Knowledge of map reading and orientation in field setting.	6.00	6.07	3.29
c. Knowledge of reconnaissance, surveillance, and target acquisition procedures.	5.50	5.93	1.43
d. Knowledge relating to communications equipment and communications procedures.	4.88	5.36	2.14
e. Communication skills (ability to use communications equipment and face-to-face communications to enhance mission accomplishment).	5.38	5.64	2.43

Appendix E. Results of Survey for System Evaluation

GROUP/SAMPLE SIZE:

A – PLT SGT / PLT LDR   = 8
B – TM LDR / SQD LDR   = 7
C – Civilian   = 9

A. Using the scale below, please rate what you feel would be the importance of the following items.

1	2	3	4	5	6	7
No experience	Below average	Slightly below average	Average	Slightly above average	Above average	Expert

	MEAN RESPONSE		
	A	**B**	**C**
Ability to Sense:			
1. Inanimate nonmilitary objects	5.63	5.17	5.13
2. IEDs	7.00	6.71	6.56
3. Weapons	6.75	6.71	6.56
4. Civilians	6.00	5.71	5.78
5. Friendly/Enemy military personnel	7.00	6.57	6.33
6. Other	6.00	6.00	7.00
Information fidelity			
7. Ability to see actual images and discern details concerning the images such as facial expressions	5.13	5.14	5.33
8. Ability identify targets (i.e., friend, civilian, or foe; M16 weapon versus AK47 weapon)	6.50	6.43	6.00
9. Ability to recognition targets (i.e., human is present, weapon is present)	6.63	6.57	6.67
10. Ability to classify information into categories (i.e., man, machine, inanimate objects, etc.)	5.75	6.00	6.33
11. Ability to detect targets (i.e., something is in that room)	6.63	6.29	6.22
Information Latency			
12. Ability to transmit information instantaneously	6.75	7.00	6.67
13. Ability to transmit information with a 5 to 10 second latency	6.25	5.00	5.11
14. Ability to transmit information with a 10 to 30 second latency	5.88	4.29	4.67
15. Ability to transmit information with a 30 second to 1 minute latency	5.50	4.00	4.33
16. Ability to return with downloadable information	6.50	3.00	3.56

(cont)	MEAN RESPONSE		
	A	B	C
Single Task/Mission Duration			
17. Duration up to 15 minutes	5.25	4.50	5.67
18. Duration up to 30 minutes	5.71	4.67	5.78
19. Duration up to 1 hour	6.00	4.83	5.56
20. Duration up to 2 hours	6.29	5.50	5.67
21. Duration up to _____ hours	6.67	7.00	6.29
Total Operational Time Before "Recharge"			
22. Up to 15 minutes	5.00	4.20	4.88
23. Up to 30 minutes	5.43	4.40	4.88
24. Up to 1 hour	5.75	4.60	5.25
25. Up to 2 hours	6.38	4.80	5.63
26. Up to _____ hours	6.86	6.86	6.17
Mobility			
27. Ability to fly	6.86	6.71	6.38
28. Ability to crawl	6.43	6.57	6.50
29. Ability to climb walls	6.57	6.57	6.37
30. Ability to maneuver through rubble	6.86	6.86	6.00
31. Ability to do all mobility maneuvers listed above	6.88	6.71	5.13
32. Ability to keep up with squad movement	7.00	6.00	5.38
33. Ability to keep up ahead of squad	6.63	6.29	5.13
System Command & Control (C2) Echelon			
34. Ability to C2 the system from the squad	6.50	6.43	6.00
35. Ability to C2 the system from the platoon	6.25	5.86	5.71
36. Ability to C2 the system from the company	6.13	5.57	4.86
37. Ability to give directions to the system controller from the squad	6.88	6.17	6.71
38. Ability to give directions to the system controller from the platoon	6.50	5.33	6.14
39. Ability to give directions to the system controller from the company	5.88	5.08	5.86
Level of Autonomy			
40. Completely autonomous system (i.e., system makes decisions concerning which direction to turn, what altitude to fly, which room to enter based upon predetermined rules)	5.38	5.71	5.38
41. System is given instructions concerning the intent of the leader and then it makes decisions on how to carry out the instructions	6.50	6.29	6.25
42. System is given waypoints and goes to those waypoints but makes decisions concerning how to get to the waypoints in the most efficient way (i.e., obstacle avoidance)	6.50	6.14	6.38

(cont)	MEAN RESPONSE		
	A	B	C
43. System is given waypoints but alerts the operator concerning any problems that arise to ask for further instruction	6.75	6.43	5.50
44. System is tele-operated and all decisions are made by the operator	5.88	4.57	3.63
Stealth			
45. Audible stealth	6.88	6.57	6.75
46. Visual stealth (day)	6.75	6.43	6.75
47. Visual stealth (night)	6.88	6.00	6.13
48. Electromagnetic stealth	6.57	6.14	4.38
49. Thermal stealth	6.38	5.86	4.50
50. Other type of stealth:	6.67	7.00	7.00
Physical Specifications			
51. Lightweight	6.75	6.57	6.57
52. Small Size	6.63	6.57	6.43
53. Accomplish task with one platform	6.13	5.57	5.29
54. Accomplish task with multiple platforms	5.63	6.86	6.00

Comments   No. of Comments

A

Q21: as long as possible.	1
Q21: 12.5 hrs (mean)	4
Q26: as long as possible.	1
Q26: 14 hrs (mean)	5

B

Q21: 48 hrs (mean)	7
Q26: 49 hrs (mean)	7
Q50: Camouflage.	1

C

Q1: for obstacle avoidance.	2
Q6: humans.	1
Q12: show instant transmit on demand.	1
Q21: 34 hrs (mean)	7
Q26: 36 hrs (mean)	7

B. Please rank order the following characteristics in terms of which ones are the MOST important capabilities needed by the MAST systems. Place a 1 next to the most important characteristic, a "2" next to the second most important characteristic, etc. There should be no ties.

MAST CTA: 10 = best, 1 = worst

Capability	Plt leaders		T/S leaders		Civilians	
	Mean	SD	Mean	SD	Mean	SD
Ability to sense	5.86	3.24	5.50	3.27	8.56	2.13
Information fidelity	7.71	2.50	8.71	1.11	6.25	1.91
Information latency	6.86	2.54	6.57	3.31	6.00	1.07
Single task/mission duration	3.00	1.91	4.33	3.01	4.38	1.69
Total operational time before recharge	5.71	3.04	5.83	2.48	3.75	1.28
Mobility	7.00	1.83	7.50	1.87	7.88	1.36
System command & control echelon	4.43	2.94	2.83	1.72	2.25	1.98
Level of autonomy	4.71	2.21	4.83	2.32	6.38	2.07
Stealth	7.00	2.58	7.57	2.07	8.89	1.96
Physical specification	2.71	2.21	2.83	2.23	1.88	1.36

C. What TTP need to be established for use of MAST systems?

Comments   **No. of Comments**

A

TTP change as the battlefield does. To say what would be a good TTP without   1
  working with the equipment itself is something I would not push just yet.
How to deploy them.   2
How to operate in MOUT environment.   1
What happens when you encounter IEDs?   1
Who operates MAST systems?   1
Depends on the capabilities of MAST and type of unit.   1
Use of the system in a given unit. This could vary from who operates them to when   1
  to use them or not use them.

B

Can only be determined once system is finalized and integrated into military tactical   1
  doctrine and practice.
System needs to be put into an exercise before you are able to build TTP around it.   2
Should be controlled by squad leader.   1

C

Rehearsal protocols.   1

Comments	No. of Comments
Command and information processing.	1
Optimal configuration/selection of systems for typical situations.	1
Use before a mission to obtain intelligence (Intel).	1
TTP likely cannot be undertaken until MAST is demonstrated.	1
Use during big clearing, if at all.	1
Non-specific recon.	1
Guidelines regarding who has access to them, who has control of them, and who sees information obtained by them.	1
The level of control of the system.	1

D. List three things you liked about the concept of a MAST system.

A

Size.	3
Small unit control.	1
Communication enhancement.	1
Capabilities.	1
Ability to detect IEDs.	1
Ability to maneuver.	1
Ability to detect personnel.	1
Able to use as reconnaissance.	2
Multiple purpose sensor.	1
Early warning.	2
Prevents Soldiers from unnecessary risks.	1
Ability to provide source of security.	1
Real time feed.	1
Ability to provide an enhanced situational environment.	1
Ability to provide another platform for information.	1

B

The idea is great.	1
Maneuverability.	1
Has no human weakness/errors.	1
It can see and hear what I can't.	1
More surveillance.	1
Recon abilities.	2
Have eyes on prior to mission.	1
Small, lightweight.	1
Size of the systems.	1
Able to move anywhere.	1
Ability to help secure areas without jeopardizing Soldiers' lives.	1
Level of technology.	1
A step in the future of tomorrow's warfare.	1
Capabilities of the systems.	1

Comments	**No. of Comments**
Helps save lives.	1
Will help with Intel and Intel gathering, pre-mission planning, and AAR.	1
Help to put us on an offensive mode more than a defensive mode.	1
Ability for new things to be generated.	1

C

Provides current information.	1
Better Intel before mission.	2
Provide perimeter security.	1
Easy to deploy.	1
Ability to be pre-deployed.	1
Minimal weight.	2
Small sizes provide indoor navigation ability.	1
Autonomy.	1
Flexibility.	1
Mobility.	1
Stealth.	4
Cost.	1
Little effort to use.	1
Mobility of sensors.	1
Negate some missions that are deemed unnecessary from Intel.	1
Has the potential to help squad-level teams.	2
Can significantly enhance situational awareness and be used to determine when an operation is not necessary (i.e., there is no target in the building) which seems to be a great problem.	1
Information available down to a low level (i.e., squad/fire team).	1
Persistent surveillance.	1

E. List three things you believe can be improved in the MAST system designs/capabilities.

A

I think we are on the right track so I have nothing to add.	1
Ability to hear.	1
Lethal capability.	1
Long-range capabilities.	1
Relay communications over long distances.	1
Stealth.	1
Sights.	1
Ability to attach hearing devices on individuals.	1
Mobility.	1
Detect explosives.	1

Comments	No. of Comments
Move through jungle environment.	1
Weapon system.	1
Speed.	1
Size.	1

B

Design so that enemy can't turn it around and use it on us.	1
Interface.	1
Time to get into play.	1
Realistic employment.	1
Ability of life of battery duration.	1
Multiple sensors, day sight, night vision sight, thermal sight, audible sensors, and RF sensors on multiple platforms.	1

C

Focus on human-system interface.	1
Must be tamper-proof or self-destruct.	1
Find ways to keep them from alerting the enemy.	1
Enemy should learn to defeat them.	1
Baby step oriented objectives.	1
Focus on 3–5 types of systems after 3 years.	1
Behaviors/swarming.	1
Stealth.	1
Duration.	1
Autonomy.	1
Command user interface/switching over control.	1
Sensor fusion and information presentation to Soldiers.	1
Information fidelity to the Warfighter will determine whether it is ever used.	2
There seems to be a great many scenarios driving the research, but all require significantly different capabilities.	1
It is not clear whether sensing can be interpreted and helpful during battle (i.e., if the camera view is bad). This needs to be addressed.	1
Sensors available.	1
HMI.	1

F. List up to three things you would like the MAST system to do that were not discussed.

A

I think we hit all the key areas.	1
Detect/destroy IEDs.	1
Distinguish between friend/foe.	1

B

Contain explosive devices.	1
Ability to self-detonate.	1
Solar powered.	1
Ability to detect moods.	1
Communicate verbally.	1
To deploy and redeploy by itself.	1
To understand this is a tool; it can't replace boots on the ground.	1

C

Fully controlled options; no autonomy.	1
Provide light in caves.	1
Provide maps of caves, etc. (automatic map generation).	1
Drop payload and leave.	1
Morphing from one system to another.	1
Combining systems into a large one (transformers).	1
Set it up to turn off after a given time and only turn on again when given a "password" so they can't be used by the bad guys.	1
Immobilize bad guys before we get there.	1
Fusion of information into a global database that can be accessed at multiple levels.	1
Ability to penetrate a building without explicit control, i.e., finding windows, etc., automatically.	1
Amount of training.	1
Quantity of MAST assets required to make a difference.	1
How system can evolve as new technologies/threats emerge.	1

G. Additional Comments:

A

We need this now, not 20 years from now. This is a very good program; please don't let it die.	1
I hope that in the future the MAST system will give us the ultimate edge to win on the battlefield.	1

B

Great idea and hope it comes out sooner than 20 years.	1
Awesome design and great people working on project.	1
Really enjoyed classes and the experience. Thank you for your time and giving us "green suiters" the time to help with the design and capabilities.	1
Outstanding concept; however, the greatest problem will be the people creating it. OPSEC … maybe this should be developed and used only by top members of the NSA, CIA, special operations forces with a minimum TOP SECRET security clearance.	1
These ideas are crazy as hell, but I like it. Keep it up.	1

Comments	No. of Comments

C

There may need to be a fundamental change in SOP as MAST systems get adopted.	1
Need to discuss level of learning.	1
Need to consider heterogeneous in terms of size, level of mobility, and operating time.	1
Baby steps are critical to developing a systems capability over a long period of time.	1
Spiral development is probably best approach.	1

Appendix F. Acronyms

Acronym	Definition
ARL	Army Research Laboratory
ASST	Assistant
C2	Command and Control
CAM	Cooperative Agreement Manager
CPT	Captain
CTA	Collaborative Technology Alliance
CTNSP	Center for Technology and National Security Policy
DIME	Diplomatic, Information Operations, Military, and Economic
DoD	Department of Defense
DTP	Defense and Technology Paper
FBCB2	Force XXI Battle Command Brigade and Below
GNR	Gunner
HRED	Human Research and Engineering Development
HUMINT	Human Intelligence
IED	Improvised Explosive Device
IMINT	Imagery Intelligence
IW	Irregular Warfare
KSA	Knowledge, Skills, and Abilities
LDR	Leader
LOO	Line of Operation
MAST	Micro-Autonomous Systems and Technology
METT-T	Mission, Enemy, Terrain, Troops, and Time Available
MG	Machine Gun
MOS	Military Occupational Specialty
NCO	Non-Commissioned Officer
NDU	National Defense University
OBJ	Objective
OP	Observation Post
OPCON	Operational Control
OPFOR	Opposing Force
OPTEMPO	Operations Tempo
OSD	Office of Secretary of Defense
PLT	Platoon
PMESII	Political, Military, Economic, Social, Information, and Infrastructure
RATELO	Radio-Telephone Operator
ROE	Rule of Engagement
S&T	Science and Technology
SA	Situational Awareness
SBF	Support by Fire
SCU	Small Combat Unit

SFC	Sergeant First Class
SGT	Sergeant
SIGINT	Signal Intelligence
SQD	Squad
SME	Subject Matter Expert
SSG	Staff Sergeant
SSW	Smart Sensor Web
SUGV	Small Unmanned Ground Vehicle
TIGR	Tactical Ground Reporting
TM	Team
TTP	Tactics, Techniques, and Procedures
UAS	Unmanned and Autonomous Systems